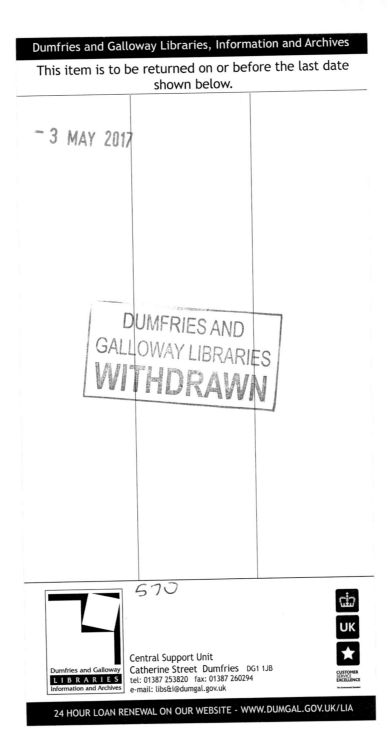

BrightRED Revision

# Higher
# HUMAN BIOLOGY

Cara Matthew

First published in 2010 by:

**Bright Red Publishing Ltd**
**6 Stafford Street**
**Edinburgh**
**EH3 7AU**

A CIP record for this book is available from the British Library

ISBN 978-1-906736-23-1

With thanks to Ken Vail Graphic Design, Cambridge (layout), Anna Clark (copy-edit)

Cover design by Caleb Rutherford – 'Eidetic'

Illustrations by Beehive Illustration (Mark Turner), Ken Vail Graphic Design

*Acknowledgements*
Bright Red Publishing and the author would like to thank Clare Little for her contribution to the text in some areas.

Every effort has been made to seek all copyright-holders. If any have been overlooked, then Bright Red Publishing will be delighted to make the necessary arrangements. All Internet links in the text were correct at time of going to press.

Permission has been sought from all relevant copyright holders and Bright Red Publishing is grateful for the use of the following:
An image showing amniocentesis. Copyrighted and used with permission of Mayo Foundation for Medical Education and Research, all rights reserved (page 40); A diagram adapted from page 346 of *Human Histology* by Alan Stevens and James Lowe, published by Mosby. Copyright Elsevier 2004 (page 45); A graph adapted from page 129 of *Illustrated Human and Social Biology* by Brian S Beckett. By permission of Oxford University Press (page 50); A graph adapted from page 24 of *AS Revision Notes Biology* published by HarperCollins Publishers Ltd © 2006 Alan Morris and Margaret Baker (page 58); A graph adapted from page 162 of *Higher Human Biology* by Torrance, published by Hodder & Stoughton 1992 (page 59); A graph adapted from page 318 of *Advanced Human Biology* published by HarperCollins Publishers Ltd © 1989 J. Simpkins and J.I. Williams (page 67); A memory graph adapted from 'Analysis of rehearsal processes in free recall' by D. Rundus taken from *Journal of Experimental Psychology*, 89, pp 36-77 (1971). Copyright 1971 The American Psychological Association. Adapted by permission (page 75).

Printed in Scotland by Thomson Litho Limited, East Kilbride, Scotland

# CONTENTS

# HIGHER HUMAN BIOLOGY
# INTRODUCTION

## COURSE STRUCTURE

The Higher Human Biology course is divided into three units:

- Unit 1: Cell function and inheritance
- Unit 2: The continuation of life
- Unit 3: Behaviour, populations and environment

## ASSESSMENT

The Higher Human Biology course is assessed in three ways:

- Each of the three units is assessed within your school using a NAB test. NABs are set by the SQA and consist of structured, short-answer questions at grade C. You must gain at least 26 marks out of a possible 40 (or 65%) to pass.

- Practical abilities are also assessed internally. You are required to write a report to a satisfactory standard on one of the investigations that you have carried out (usually from Unit 1).

- You will also take an externally-assessed written examination consisting of a paper lasting 2·5 hours. The examination has an allocation of 130 marks and is divided into three sections:

**1** Section A is worth 30 marks and consists of 30 multiple-choice questions; 20 of these test knowledge and understanding, and ten test problem-solving and practical abilities.

**2** Section B contains structured questions and is worth 80 marks; 50 marks are designed to test knowledge and understanding, and 30 marks test problem-solving and practical abilities.

**3** Section C contains two essay (or extended-response) questions, each marked out of ten. Within each question, you will have to choose between two essay titles. The first is a structured essay that is divided into parts, with the marks for each part being indicated. The second essay is open-ended and carries one mark for coherence and one mark for relevance.

### Exam hints

You do not need to answer the questions in order. Find a question that you can answer easily, so that you settle your nerves and relax a bit.

Keep an eye on time. As a general rule, you should be taking about one minute per mark. So, allowing ten minutes for settling at the start and checking your paper at the end, the timing for each section should be roughly:

- multiple-choice questions – 30 minutes
- structured questions – 1 hour 20 minutes
- each essay question – 15 minutes

The course award is graded A, B, C or D depending on how well you do in the external examination. In order to gain the course award, you must also pass the three NABs (one for each unit) and complete the investigation report to the standard required by the SQA.

## THE STRUCTURE AND AIM OF THIS BOOK

There is no short-cut to passing any course at Higher level. To obtain a good pass requires consistent, regular revision over the duration of the course. The aim of this revision book is to help you to achieve this success by providing you with a concise and engaging coverage of the Higher Human Biology course material. We recommend that you use this book, in conjunction with your class notes, to revise each topic area, prepare for NABs and prelims, and in your preparation for the final exam.

The book is divided into a section for each of the three units of the course. Within each section, there is a double-page spread on each of the sub-sections.

Each double-page spread:

- provides the key ideas and concepts of the sub-section in a logical and digestible manner.

- contains 'Don't forget' boxes that flag up vital pieces of knowledge that you need to remember and important things that you must be able to do.

- contains a 'Let's think about this' feature which will extend your knowledge and understanding of the subject, and provide additional interest. Sometimes there are questions to help you check your understanding.

## REVISION TIPS

- Don't leave your revision until the last minute. Make up a revision schedule and stick to it. Be realistic – you should work around your other activities and remember that you do need to take time off to relax.

- Find somewhere to study that is quiet and uncluttered. You need space to work.

- Study for short periods (between 30 and 45 minutes) with short breaks in between. This will help you to concentrate. During each break, go out of the room where you are studying and come back refreshed for your next study session.

- Read over each sub-topic at a slower pace than you would usually do (so that you have time to develop a good understanding), and ask yourself questions or read out loud.

- It's often easier to remember facts if you talk about topics with a family member or a friend. So, find a study buddy who can ask you questions about your work.

- Practice makes perfect; do past-paper practice so that the exam format is as familiar as possible. There are only a few ways in which you can be asked the same question – and you will see similar questions cropping up over many past papers.

- In the run-up to the exams, make sure that you get plenty of sleep, and eat lots of fresh fruit and vegetables to keep your energy levels up.

Good luck, and have fun!

# ENZYMES I

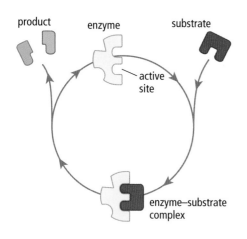

product
enzyme
substrate
active site
enzyme–substrate complex

Enzymes are **globular proteins** that act as **biological catalysts**, both inside and outside the cell. They speed up chemical reactions without being used up in the reaction. To do this, enzymes lower the activation energy – the energy needed to start a reaction. At normal body temperature, reactions would take place at too slow a rate to maintain life if enzymes were not present.

The substance upon which an enzyme acts is called a **substrate**. Each enzyme can only act on one substrate; enzyme action is, therefore, said to be **specific**. This is because the part of the enzyme that binds with the substrate (the **active site**) has a shape which is opposite to the shape of the substrate, allowing the two molecules to fit together like a lock and key. At the end of the reaction, a **product** separates from the enzyme.

## FACTORS THAT AFFECT ENZYME ACTIVITY

### Temperature

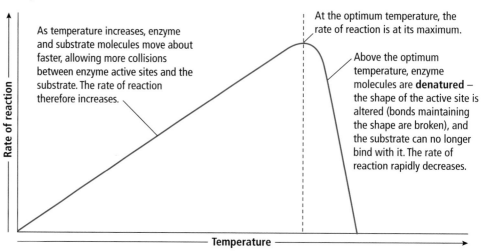

As temperature increases, enzyme and substrate molecules move about faster, allowing more collisions between enzyme active sites and the substrate. The rate of reaction therefore increases.

At the optimum temperature, the rate of reaction is at its maximum.

Above the optimum temperature, enzyme molecules are **denatured** – the shape of the active site is altered (bonds maintaining the shape are broken), and the substrate can no longer bind with it. The rate of reaction rapidly decreases.

Rate of reaction

Temperature

### pH

Every enzyme is able to catalyse reactions within a working range of pH. Within this range, the **optimum pH** is the pH at which the rate of reaction is greatest. Extremes of pH cause bonds to break, and the shape of the active site is altered; that is, the enzyme is denatured.

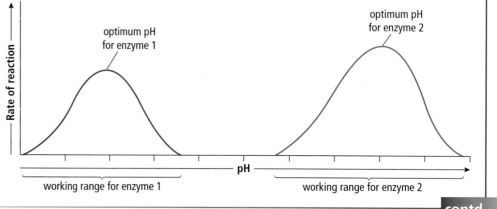

optimum pH for enzyme 1

optimum pH for enzyme 2

Rate of reaction

pH

working range for enzyme 1

working range for enzyme 2

contd

## FACTORS THAT AFFECT ENZYME ACTIVITY contd

### Enzyme concentration

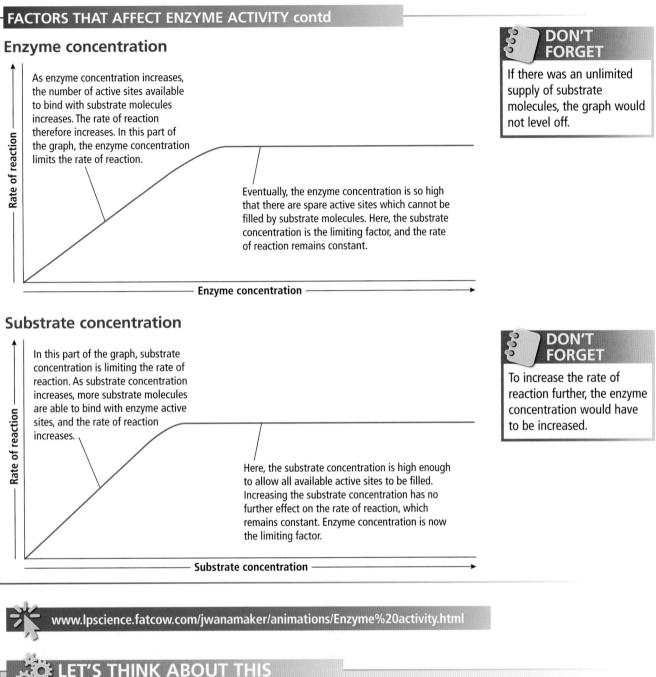

As enzyme concentration increases, the number of active sites available to bind with substrate molecules increases. The rate of reaction therefore increases. In this part of the graph, the enzyme concentration limits the rate of reaction.

Eventually, the enzyme concentration is so high that there are spare active sites which cannot be filled by substrate molecules. Here, the substrate concentration is the limiting factor, and the rate of reaction remains constant.

**Rate of reaction** (y-axis)

**Enzyme concentration** (x-axis)

### Substrate concentration

In this part of the graph, substrate concentration is limiting the rate of reaction. As substrate concentration increases, more substrate molecules are able to bind with enzyme active sites, and the rate of reaction increases.

Here, the substrate concentration is high enough to allow all available active sites to be filled. Increasing the substrate concentration has no further effect on the rate of reaction, which remains constant. Enzyme concentration is now the limiting factor.

**Rate of reaction** (y-axis)

**Substrate concentration** (x-axis)

> **DON'T FORGET**
>
> If there was an unlimited supply of substrate molecules, the graph would not level off.

> **DON'T FORGET**
>
> To increase the rate of reaction further, the enzyme concentration would have to be increased.

www.lpscience.fatcow.com/jwanamaker/animations/Enzyme%20activity.html

### LET'S THINK ABOUT THIS

1. Which part of an enzyme molecule binds with the substrate?
2. Why is enzyme action said to be specific?
3. Explain why enzymes are denatured at high temperatures.
4. In the graph, identify the limiting factor at points X and Y. Describe how the rate of reaction could be increased further.

Candidates who have not studied Standard Grade Biology might find it useful to research the substrate and product for each of the following enzymes: amylase, pepsin, maltase, lipase, catalase.

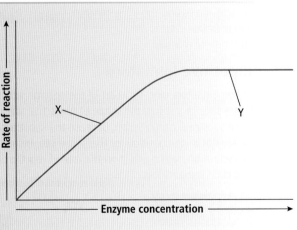

**Rate of reaction** (y-axis)

**Enzyme concentration** (x-axis)

# ENZYMES II

## ENZYME ACTIVATORS

Some enzymes are produced in an inactive form and are activated either by the presence of cofactors or by other enzymes.

### Cofactors

Cofactors are non-protein molecules which act by altering the shape of an enzyme's active site, allowing the substrate to bind with the enzyme. Mineral ions such as copper, magnesium and zinc can act as cofactors. Other cofactors are organic molecules (**coenzymes**) which often contain vitamins, such as vitamin $B_{12}$.

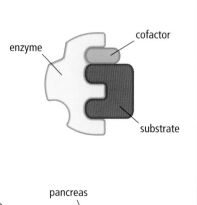

### Enzymes

Digestive enzymes are often made in an inactive form so that they do not digest the glands in which they are produced. Once released into the gut tube, the actions of an **enzyme activator** convert the inactive enzyme into its active form.

You may be familiar with the activation of a protein-digesting enzyme (a protease) called **trypsinogen**. The enzyme is produced in an inactive form in the pancreas and becomes activated within the small intestine by the reaction shown opposite.

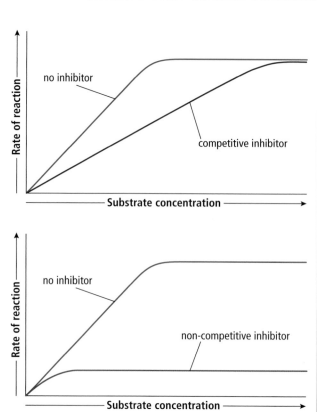

## INHIBITORS

### Competitive inhibitors

Competitive inhibitor molecules have a shape similar to the substrate. They bind with the active site, preventing the substrate from entering. Because the substrate and inhibitor are in competition for the active site, increasing the substrate concentration causes an increase in the rate of reaction.

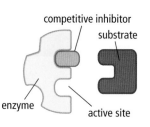

### Non-competitive inhibitors

Non-competitive inhibitors bind to a part of the enzyme which is not the active site. As a result, the shape of the active site is altered and the substrate cannot enter. Because the substrate and inhibitor are not in competition for the active site, increasing substrate concentration has no effect on the rate of reaction. The rate of reaction remains low.

**DON'T FORGET**

**Cyanide** is a non-competitive inhibitor that inhibits aerobic respiration.

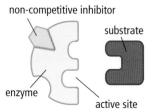

## INBORN ERRORS OF METABOLISM

All the chemical reactions that take place in an organism make up the **metabolism**.

A **metabolic pathway** is a series of chemical reactions that follow on, one after another. Each stage in the pathway is controlled by an enzyme, with the product of one reaction becoming the substrate for the next reaction.

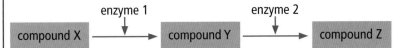

In the diagram above, compound Z can only be produced if both enzyme 1 and enzyme 2 are present.

Each enzyme is coded for by a different gene. When the base sequence of one gene is faulty (see Gene Mutations, page 38), the enzyme is not produced and the metabolic pathway will be blocked.

This is called an **inborn error of metabolism**. Inborn errors of metabolism occur when the individual is homozygous for the faulty gene. In the above example, if enzyme 2 is not present, compound Y builds up and compound Z is not produced.

### Phenylketonuria (PKU)

**PKU** is an example of an inborn error of metabolism, where absence of a specific enzyme in a metabolic pathway results in failure to break down the amino acid **phenylalanine** (taken in through the diet) and its accumulation in the body. Sufferers inherit two copies of a mutated gene.

The normal metabolic pathway is shown below.

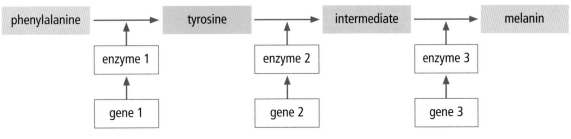

In sufferers of PKU, gene 1 is mutated and enzyme 1 is absent. The metabolic pathway is blocked, preventing production of tyrosine and melanin.

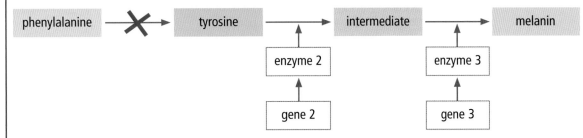

With the normal pathway blocked, phenylalanine is broken down into toxic substances that affect brain cells, causing mental disability and seizures. PKU can be controlled by eating a diet low in phenylalanine.

Look up www.ygyh.org/pku/cause.htm

## LET'S THINK ABOUT THIS

Sufferers of PKU have very fair skin and blue eyes, but are not albinos. Why?

Although phenylalanine cannot be broken down in PKU, some tyrosine forms part of the normal diet. This tyrosine can be broken down to produce melanin, which gives some pigmentation to the skin.

# PROTEIN STRUCTURE AND FUNCTION

## STRUCTURE OF PROTEINS

All proteins contain the elements **carbon**, **hydrogen**, **oxygen** and **nitrogen**. These elements combine to form an **amino acid**, the basic unit of a protein. There are approximately 20 naturally-occurring amino acids, giving rise to a huge variety of proteins.

### Primary structure

Amino acids are brought together in a pre-determined order to form a chain called a **polypeptide**. In the chain, amino acids are held together by **peptide bonds**.

amino acid

peptide bond

### Secondary structure

Once the primary structure has formed, the polypeptide becomes coiled and is held together by hydrogen bonds.

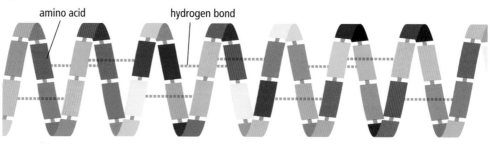

amino acid          hydrogen bond

### Tertiary structure

Finally, the polypeptide forms sheets (fibrous proteins) or is wound up to form a ball (globular proteins). Each protein's shape is maintained by bonds.

polypeptide                polypeptide

fibrous proteins – flat sheets          globular proteins – wound into a ball

## VARIETY OF PROTEINS

| | **Fibrous proteins** |
| --- | --- |
| | Long parallel chains of polypeptides are held together by cross-linkages that make the structure very strong, like a rope. These are used in structural proteins in tissues. For example, collagen provides strength in tendons and bones; actin and myosin are the contractile proteins in muscle cells. |

**contd**

## VARIETY OF PROTEINS contd

| Globular proteins | | |
|---|---|---|
| Polypeptide chains are wound into a ball (like a tangled ball of wool), with the shape maintained by weak bonds. | | |
| **Type of globular protein** | **Example** | **Function** |
| (a) structural | cell membrane proteins | Membrane proteins are involved in several important processes in the membrane, including support. |
| (b) enzymes | amylase | Shape provides an active site into which a substrate fits. |
| (c) hormones | insulin | Chemical messengers that target specific tissues to exert an effect. |
| (d) antibodies | | Y-shaped proteins that provide two binding sites for antigens. |
| (e) transport proteins | haemoglobin | Haemoglobin is composed of protein bound to iron. It is involved in transport of oxygen. |

Look up http://biology.about.com/od/molecularbiology/a/aa101904a.htm

## LET'S THINK ABOUT THIS

The diagram below shows the arrangement of protein fibres in skeletal muscle. You must be able to:
- identify actin and myosin filaments
- explain how contraction of skeletal muscle is achieved.

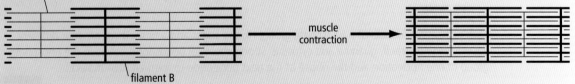

During muscle contraction, filament A (actin) and filament B (myosin) slide over each other, causing the overall length of the muscle to decrease. Note that actin and myosin filaments do not change in length.

http://course1.winona.edu/sberg/ANIMTNS/SlidFila.htm

You may be asked to write an essay on the functions of proteins. Use your notes to write a bullet-point list under the following headings: enzymes, hormones, muscular contraction, transport, antibodies, and structural proteins.

# NUCLEIC ACIDS

## DNA

**Deoxyribonucleic acid** (**DNA**) is a double-stranded molecule made up of subunits called **nucleotides**.

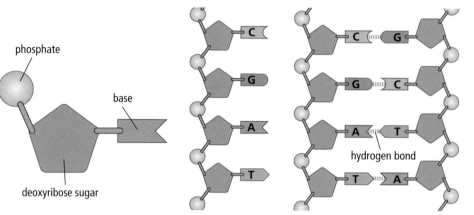

Each nucleotide consists of a **phosphate** group, **deoxyribose** sugar and a **base**. The four different types of nucleotide each have a different base: **adenine**, **thymine**, **guanine** and **cytosine**. A chemical bond forms between the phosphate group of one nucleotide and the deoxyribose sugar of the next nucleotide, producing a strong strand.

Complementary base pairing occurs between the bases of two DNA strands. The bases are held together by hydrogen bonds, giving a double-stranded DNA molecule which then twists to form a double helix. In DNA, adenine always bonds with thymine (**A–T**), and cytosine always bonds with guanine (**C–G**).

**www.zerobio.com/mendel1a.htm**

## RNA

Ribonucleic acid (RNA) is a single-stranded molecule made of nucleotides.

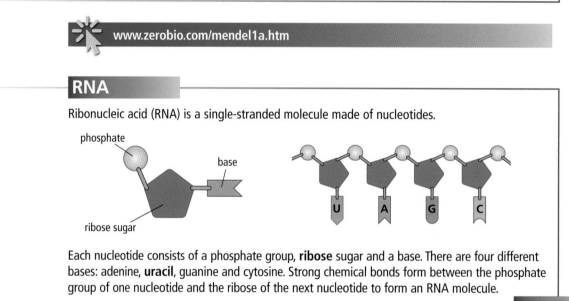

Each nucleotide consists of a phosphate group, **ribose** sugar and a base. There are four different bases: adenine, **uracil**, guanine and cytosine. Strong chemical bonds form between the phosphate group of one nucleotide and the ribose of the next nucleotide to form an RNA molecule.

**contd**

## RNA contd

You should be familiar with the following types of RNA.

### mRNA

Messenger RNA is made in the nucleus by transcription of DNA (see page 14). It passes through the nuclear pores to enter the cytoplasm, where it travels to the ribosomes. mRNA is involved in protein synthesis. On mRNA, a triplet of bases (a **codon**) codes for one amino acid.

### tRNA

Transfer RNA is made in the nucleolus inside the nucleus, and leaves through the pores to reach the cytoplasm. Here, each tRNA molecule binds temporarily with an amino acid, carrying it to a ribosome for protein synthesis. Each type of amino acid attaches to a different type of tRNA. As there are approximately 20 different amino acids, there are approximately 20 different types of tRNA. tRNA is a clover-shaped molecule with an exposed triplet of bases called the **anticodon**.

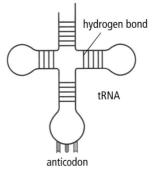

hydrogen bond

tRNA

anticodon

### rRNA

Ribosomal RNA is made in the nucleolus and passes out of the nucleus to form part of the structure of a ribosome.

## LET'S THINK ABOUT THIS

You should be able to calculate how many molecules of an individual base are on a DNA strand when given the number of amino acids in the polypeptide and the percentage composition of any other base within the strand. Have a look at the example below.

How many guanine bases are on a DNA strand that codes for a polypeptide chain 300 amino acids long, if 20% of the DNA strand is adenine?

300 amino acids are coded for by 900 bases. If 20% of the strand is adenine, then 20% must be thymine, as they form complementary base pairs. The remaining 60% are cytosine–guanine pairs. Therefore, 30% of 900 bases are guanine = 270 bases.

Complete the following table to compare and contrast DNA and mRNA.

| | DNA | mRNA |
|---|---|---|
| Type of sugar present | | |
| Number of strands of nucleotides in molecule | | |
| Bases present | | |
| Where molecule is found in the cell | | |

# PROTEIN SYNTHESIS I

## TRANSCRIPTION

The first stage of protein synthesis takes place in the nucleus and is called **transcription**. An mRNA molecule is produced that carries the genetic code from the DNA in the nucleus to a ribosome in the cytoplasm. Production of mRNA is essential, as DNA is too large to pass through the nuclear membrane.

The process is as follows:

1 The section of a DNA molecule carrying the code for the protein to be transcribed unwinds and 'unzips', exposing the bases.

2 mRNA nucleotides move in and form complementary base pairs with one of the DNA strands (the coding strand). Weak hydrogen bonds form. Cytosine always pairs with guanine; adenine on DNA pairs with uracil on mRNA, and thymine on DNA pairs with adenine on mRNA.

3 Strong chemical bonds form between the phosphate of one nucleotide and the ribose of the next nucleotide.

4 The weak hydrogen bonds that were holding the DNA and mRNA strands together break, allowing the mRNA to leave the nucleus and enter the cytoplasm.

5 Hydrogen bonds reform between the two DNA strands, and the DNA molecule rewinds to form a double helix.

**DON'T FORGET**

A triplet of bases on mRNA is called a **codon**, and codes for one amino acid.

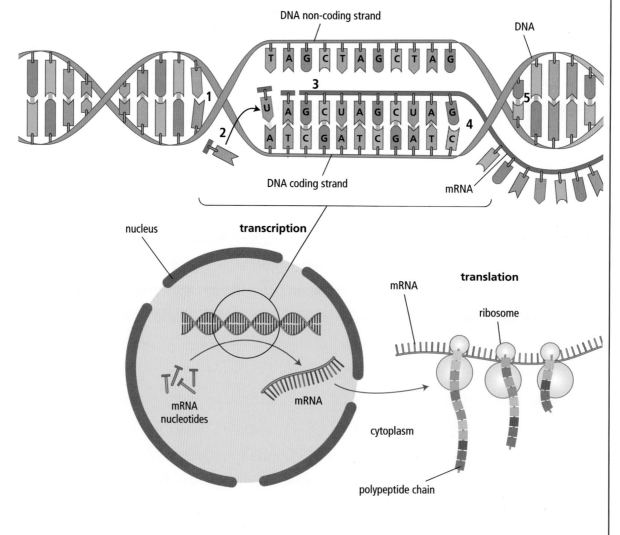

## TRANSLATION

The mRNA molecule formed during transcription becomes attached to a ribosome, either lying freely in the cytoplasm or attached to the rough endoplasmic reticulum (RER). This is where **translation** occurs, in which the order of bases on the mRNA determines the order of amino acids in a specific protein. The process is as follows:

1 **Transfer RNA** (**tRNA**) molecules become attached to amino acid molecules in the cytoplasm. Each type of amino acid attaches to a specific tRNA molecule. As there are approximately 20 different types of amino acid, there are also approximately 20 different types of tRNA. Each type of tRNA molecule has an exposed triplet of bases called an **anticodon**.

2 Transfer RNA (tRNA) molecules in the cytoplasm transport amino acids to the ribosome.

3 The first tRNA molecule moves in to allow base pairing between the triplet of bases (anticodon) on the tRNA molecule and a triplet of bases (codon) on the mRNA strand. The codon and anticodon must be complementary to each other.

4 Another tRNA molecule carries an amino acid to the ribosome. Complementary pairing between codon and anticodon brings the amino acids in line beside each other. A peptide bond forms between the amino acids.

5 Once peptide bonds have formed between the amino acids, their position is fixed. The first tRNA molecule detaches from the mRNA and is free to collect another amino acid from the cytoplasm.

6 As translation progresses, the ribosome moves along the mRNA molecule (like a zipper moves along a zip) exposing the third codon, allowing a third tRNA molecule to bring a third amino acid into position.

7 This process is repeated until the end of the mRNA strand, when a polypeptide has been formed.

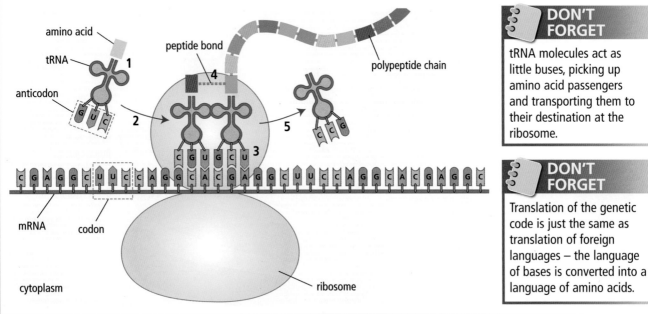

> **DON'T FORGET**
>
> tRNA molecules act as little buses, picking up amino acid passengers and transporting them to their destination at the ribosome.

> **DON'T FORGET**
>
> Translation of the genetic code is just the same as translation of foreign languages – the language of bases is converted into a language of amino acids.

Look up www.biotopics.co.uk/genes/trans.html
Look up www.artesanto.co.uk

## LET'S THINK ABOUT THIS

Use the following website to practise transcription and translation of mRNA. You must know the complementary base pairs of DNA with mRNA, and mRNA with tRNA.
http://learn.genetics.utah.edu/content/begin/dna/transcribe

# PROTEIN SYNTHESIS II

## RER AND GOLGI APPARATUS

Two cell organelles are of particular importance in protein synthesis:

- The **rough endoplasmic reticulum (RER)** consists of a series of flattened membrane sacs, continuous with the nuclear membrane and with ribosomes attached to the outer membrane surface. The RER transports newly-synthesised proteins to the Golgi apparatus.

- The **Golgi apparatus** is composed of a stack of between three and seven flattened membrane sacs. Each region of the Golgi apparatus contains enzymes that alter newly-synthesised proteins by either adding bits on or chopping bits off; vesicles move between the sacs to transfer the new proteins.

## PROCESSING AND PACKAGING

Once amino acids have been assembled into a polypeptide chain at a ribosome, the polypeptide may pass into the RER and then subsequently to the Golgi apparatus for further processing and packaging. The diagram shows the sequence of events as a protein is prepared for secretion out of a cell.

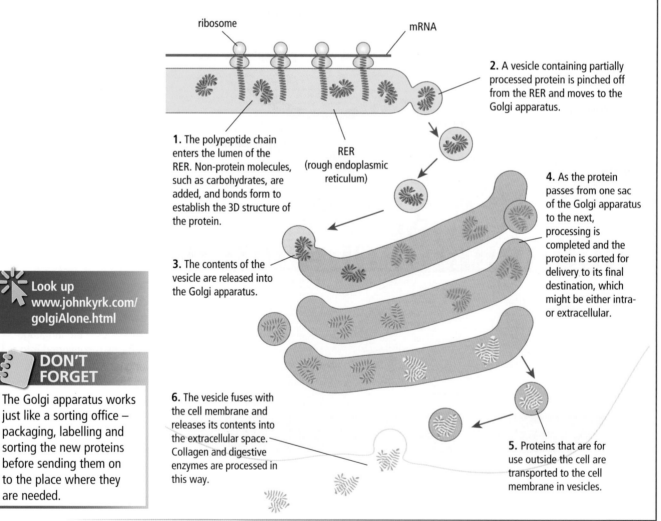

ribosome

mRNA

**2.** A vesicle containing partially processed protein is pinched off from the RER and moves to the Golgi apparatus.

**1.** The polypeptide chain enters the lumen of the RER. Non-protein molecules, such as carbohydrates, are added, and bonds form to establish the 3D structure of the protein.

RER (rough endoplasmic reticulum)

**4.** As the protein passes from one sac of the Golgi apparatus to the next, processing is completed and the protein is sorted for delivery to its final destination, which might be either intra- or extracellular.

**3.** The contents of the vesicle are released into the Golgi apparatus.

**6.** The vesicle fuses with the cell membrane and releases its contents into the extracellular space. Collagen and digestive enzymes are processed in this way.

**5.** Proteins that are for use outside the cell are transported to the cell membrane in vesicles.

**Look up**
www.johnkyrk.com/
golgiAlone.html

### DON'T FORGET

The Golgi apparatus works just like a sorting office – packaging, labelling and sorting the new proteins before sending them on to the place where they are needed.

## SUMMARY OF PROTEIN SYNTHESIS

mRNA nucleotides are assembled using a DNA template. The language of the code as a sequence of bases on DNA is retained as a sequence of bases on mRNA. It is simply transcribed (just like copying this paragraph onto another sheet of paper).

Proteins are passed through RER and on to the Golgi apparatus, where they are packaged before use and may have non-protein molecules such as carbohydrates added to them. Vesicles transport the protein between RER and Golgi apparatus.

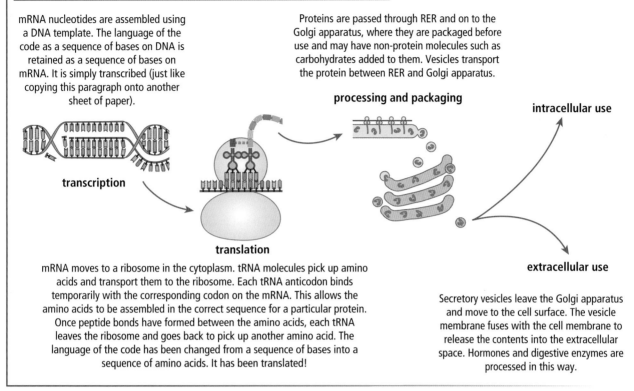

**processing and packaging**

**intracellular use**

**transcription**

**translation**

mRNA moves to a ribosome in the cytoplasm. tRNA molecules pick up amino acids and transport them to the ribosome. Each tRNA anticodon binds temporarily with the corresponding codon on the mRNA. This allows the amino acids to be assembled in the correct sequence for a particular protein. Once peptide bonds have formed between the amino acids, each tRNA leaves the ribosome and goes back to pick up another amino acid. The language of the code has been changed from a sequence of bases into a sequence of amino acids. It has been translated!

**extracellular use**

Secretory vesicles leave the Golgi apparatus and move to the cell surface. The vesicle membrane fuses with the cell membrane to release the contents into the extracellular space. Hormones and digestive enzymes are processed in this way.

## LET'S THINK ABOUT THIS

When given appropriate information, you should be able to work out corresponding sequences of DNA, RNA and peptide chains. The table below shows the mRNA codons for some amino acids.

| First position | Second position | | | | Third position |
|---|---|---|---|---|---|
| | **A** | **U** | **G** | **C** | |
| **A** | lycine asparagine | methionine/start isoleucine | arginine serine | threonine threonine | G U |
| **C** | glutamine histidine | leucine leucine | arginine arginine | proline proline | G U |

**(a)** A section of DNA has the following bases:

| T | A | C | G | T | A | G | C | C | T | C | A |

Identify the codons of the mRNA and the anticodons of the tRNA that would be used in protein synthesis from this section of DNA. Use the table of mRNA codons to identify the amino acid sequence that would be produced.

**(b)** From the table, what are the mRNA codons for (i) lycine and (ii) isoleucine?

**(c)** What is the tRNA anticodon for the amino acid arginine?

## LET'S THINK ABOUT THIS

Ribosomes that assemble proteins for use outside the cell are always attached to the RER. However, ribosomes can also be found free within the cytoplasm. Groups of ribosomes become attached to one mRNA strand, all translating the code at the same time, very quickly producing many copies of the same polypeptide for use inside the cell.

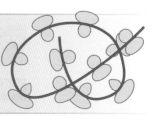

# RESPIRATION

Respiration is a series of enzyme-controlled reactions in which energy is released from food molecules. It occurs in all living cells.

## RESPIRATORY SUBSTRATES AND USES OF ENERGY

Molecules that can be broken down to release energy in respiration are called **respiratory substrates**. Glucose is the main respiratory substrate. The energy released is used to fuel cellular processes such as protein synthesis, muscle contraction, active transport and DNA replication.

### ATP and ADP

The series of reactions that make up respiration result in chemical energy being transferred to a molecule called **ATP**, adenosine triphosphate. ATP is a source of energy that can be used immediately by cells. During respiration, ATP is made when a bond forms between an **inorganic phosphate ($P_i$)** and **ADP**, adenosine diphosphate.

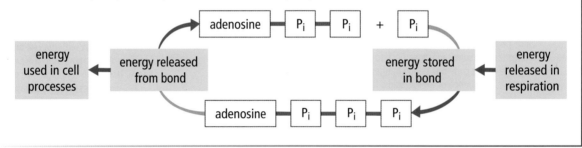

>
> **DON'T FORGET**
>
> The total mass of ATP in a cell remains relatively constant as it is made when required and used almost immediately.

## STAGE 1 – GLYCOLYSIS

Glycolysis takes place in the cytoplasm of every living cell. No oxygen is required. The conversion of 2ATP → 2ADP + $2P_i$ provides enough energy to begin the breakdown of glucose molecules. One (6C) **glucose** molecule is broken down into two (3C) **pyruvic acid** molecules. As 4ATP molecules are produced during the reaction, there is a net gain of 2ATP (4ATP produced – 2ATP used). Hydrogen is also released and picked up by a hydrogen carrier called **NAD** to make **NADH$_2$**.

> **DON'T FORGET**
>
> Anaerobic respiration in animals is a reversible reaction. Lactic acid is converted back to pyruvic acid if the oxygen debt is repaid.

> **DON'T FORGET**
>
> Glycolysis results in a net gain of two ATPs.

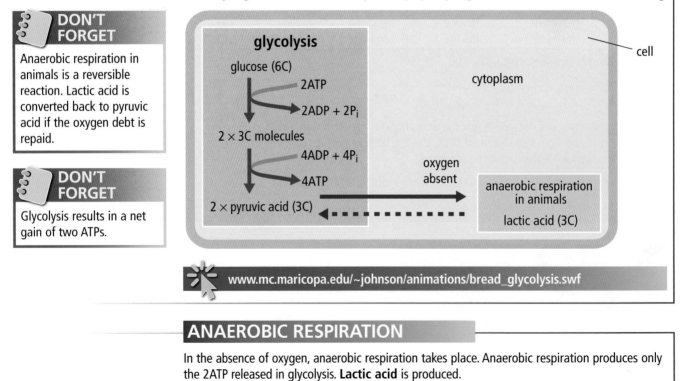

> www.mc.maricopa.edu/~johnson/animations/bread_glycolysis.swf

## ANAEROBIC RESPIRATION

In the absence of oxygen, anaerobic respiration takes place. Anaerobic respiration produces only the 2ATP released in glycolysis. **Lactic acid** is produced.

## STAGE 2 – KREBS CYCLE

On entering the **matrix of a mitochondrion**, pyruvic acid is broken down to produce **acetyl (2C) co-enzyme A**. This enters the cycle by combining with a 4-carbon molecule to produce **citric acid (6C)**. As the cycle proceeds, carbon atoms are released and combine with oxygen to form **carbon dioxide**, and hydrogen ions are released and picked up by **NAD** to form **NADH$_2$**. NADH$_2$ will carry the hydrogen ions to the third stage of respiration, the cytochrome system, on the inner membrane of the mitochondrion.

> http://www.sumanasinc.com/webcontent/animations/content/cellularrespiration.html

> http://highered.mcgraw-hill.com/sites/0072507470/student_view0/chapter25/animation__how_the_krebs_cycle_works__quiz_1_.html

**DON'T FORGET**

Every time a carbon atom is lost, $CO_2$ is produced.

**DON'T FORGET**

For each glucose molecule, Krebs cycle will go round twice – once for each of the pyruvic acid molecules produced in glycolysis.

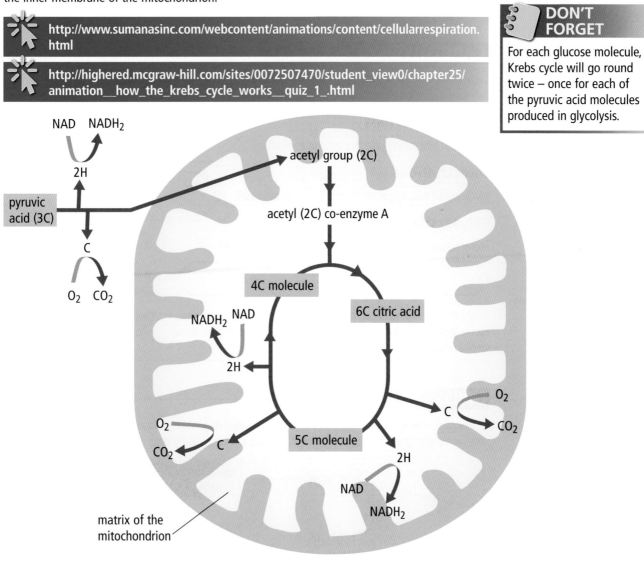

## STAGE 3 – CYTOCHROME SYSTEM

The cytochrome system is found on the **cristae** of the mitochondria and is the main site of ATP production. Hydrogen is transferred to the cytochrome system by NADH$_2$. It is then passed along a series of hydrogen carriers. This is an **aerobic process** as the final hydrogen acceptor is oxygen, which binds with hydrogen to produce **water** as a waste product. Every time a pair of hydrogen ions passes through the cytochrome system, 3ATP are produced. **36ATP** is made in the cytochrome system.

NAD ← | cytochrome system of hydrogen carriers | → oxygen
NADH$_2$ → | | → water

$3ADP + P_i$ → $3ATP$

## LET'S THINK ABOUT THIS

Use your class notes to describe the number and appearance of mitochondria in (**i**) active and (**ii**) inactive cells.

# ALTERNATIVE SOURCES OF ENERGY

Although carbohydrates are the main respiratory substrate, the body can use lipids and proteins when carbohydrates are not available.

## CARBOHYDRATES

Carbohydrates can be divided into three groups, based on the number of sugar units that each molecule contains.

### Monosaccharides

Consist of one sugar unit (for example, glucose).

### Disaccharides

Made up of two sugar units (for example, maltose).

### Polysaccharides

Many sugar units joined together (for example, starch).

Only monosaccharides can be used directly as a source of energy. When monosaccharides are unavailable, disaccharides and polysaccharides are broken down to produce monosaccharides for respiration.

## LIPIDS

Fat contains twice the energy of either carbohydrates or proteins. It can be broken down into fatty acids and glycerol, both of which can enter the respiratory pathway.

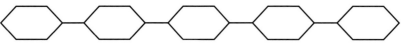

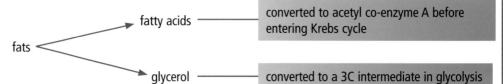

fats → fatty acids → converted to acetyl co-enzyme A before entering Krebs cycle

fats → glycerol → converted to a 3C intermediate in glycolysis

Other uses of fats in the body are shown below.

### Heat insulation

A layer of subcutaneous fat in the skin helps to insulate the body, reducing heat loss.

### Nerve insulation

Nerve fibres are surrounded by a myelin sheath which is largely made up of lipid. The myelin sheath is important as it is an electrical insulator, increasing the speed at which nerve impulses are transmitted (see page 69).

### Fat pads

Fat pads are used to protect some bones and body organs. For example, the kidneys are surrounded by a protective layer of fat which cushions them against damage. Fat pads are found in both the hands and the feet.

### Vitamin transport

Some vitamins are fat-soluble (vitamins A, D, E and K). Together with fat, they are absorbed through the gut wall and enter the lacteals and lymphatic system.

### Hormones

Some hormones (for example, oestrogen and testosterone) are steroids. Steroids are a complex type of lipid.

## PROTEINS

While excess dietary protein can be used as an energy source, most of the protein that is taken in is used for growth and repair of body tissues.

Proteins are broken down by enzymes such as pepsin and trypsin to produce amino acids. Excess amino acids are removed from the body by a process called deamination (see page 60). Some products of deamination can enter the respiratory pathway.

proteins ⟶ amino acids ⟶ either converted to pyruvic acid, acetyl co-enzyme A or a 4C molecule in Krebs cycle

## STARVATION

There are three stages in starvation.

### Initial stage

The body's glycogen store is used up.

### Adaptation stage

Fats are broken down.

### Terminal stage

Prolonged starvation results in breakdown of proteins from muscles and other tissues to produce amino acids; heart muscle failure occurs when the body's proteins have dropped to half their normal level.

## MARATHON RUNNING

Three sources of energy are used during a marathon run.

### Muscle glycogen

Initially, muscle glycogen is converted to glucose and used as an energy source; this source is exhausted after several minutes.

### Liver glycogen

Glycogen is converted to glucose, which travels to the muscles in the blood.

### Fats

As the race proceeds and glycogen stores are used up, fats are broken down to release fatty acids; these are transported in the blood to the muscles, where they enter the respiratory pathway.

Transport of fatty acids to the muscles is relatively slow; runners sometimes say they 'hit the wall' when the first two sources of energy are used up. To help combat this, runners often consume high-carbohydrate meals before the race to build up their glycogen stores, and take glucose drinks during the race.

##  LET'S THINK ABOUT THIS

Use your class notes to review how to use a key to identify carbohydrates.

# CELL TRANSPORT

## FLUID MOSAIC MODEL

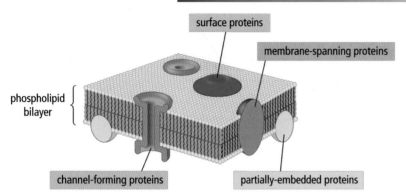

surface proteins

membrane-spanning proteins

phospholipid bilayer

channel-forming proteins

partially-embedded proteins

The plasma membrane is made up mainly of a double layer (a **bilayer**) of constantly moving (fluid) **phospholipid** molecules. **Proteins** of varying size are found scattered on and within the membrane, forming a mosaic pattern. The phospholipid bilayer allows fat-soluble molecules to pass across the membrane. The functions of some membrane proteins are given below.

### Transport proteins

Membrane-spanning (transmembrane) proteins are often involved with movement of molecules across the membrane. Channel-forming proteins create pores that allow molecules to move passively through the membrane. Other proteins use ATP as a source of energy for active transport of substances.

### Enzymes

Some proteins act as enzymes with their active sites exposed on one side of the membrane.

### Receptor molecules

Surface and partially-embedded proteins may act as receptor sites for hormones, allowing messages to be relayed into the cell.

### Antigens

Surface proteins act as antigens, allowing the cell to be recognised as 'self'.

The plasma membrane is said to be **selectively permeable**, as small molecules (such as oxygen, carbon dioxide and water) pass through freely while larger molecules, such as glucose, amino acids and urea, move through more slowly. Even larger molecules, for example proteins, are unable to pass through.

## DIFFUSION

Diffusion is the movement of molecules or ions from a region of high concentration to a region of low concentration down a **concentration gradient**. It does not use energy.

www.educypedia.be/education/biologycelldiffusion.htm
www.bcscience.com/bc8/pgs/quiz_section1.3.htm

## OSMOSIS

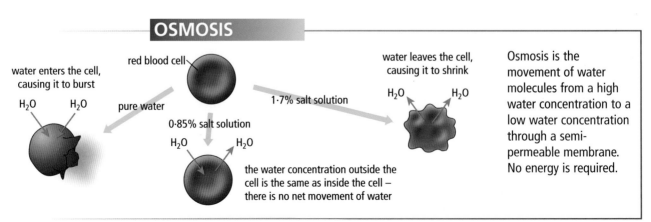

red blood cell

water enters the cell, causing it to burst

$H_2O$   $H_2O$

pure water

0·85% salt solution

$H_2O$   $H_2O$

the water concentration outside the cell is the same as inside the cell – there is no net movement of water

1·7% salt solution

water leaves the cell, causing it to shrink

$H_2O$   $H_2O$

Osmosis is the movement of water molecules from a high water concentration to a low water concentration through a semi-permeable membrane. No energy is required.

## ACTIVE TRANSPORT

**Active transport** is the movement of ions and molecules across the cell membrane **against a concentration gradient**. Molecules are moved across the membrane from a low to a high concentration by carrier proteins. As energy is required for this process, **ATP** (produced in aerobic respiration) must be available. Therefore, any factor that affects a cell's ability to produce ATP also affects the rate of active transport. These factors include: glucose concentration, oxygen concentration and temperature.

> **DON'T FORGET**
>
> Temperature affects enzyme-driven reactions such as respiration (required for energy release in active transport). As temperature increases, the rate of reaction increases until the enzyme is denatured.

## ENDOCYTOSIS AND EXOCYTOSIS

Sometimes substances move into or out of the cell through the gross movement of the cell membrane, rather than by passing directly through the membrane. Materials from the extracellular space become incorporated into the cell by the process of **endocytosis**. Secretion of materials from the cell into the extracellular space occurs by **exocytosis**.

### Endocytosis

Endocytosis includes the movement of solid materials (**phagocytosis**) and fluids (**pinocytosis**). In both cases, the process is the same: the cell membrane surrounds the material, forming a membrane-bound vesicle that moves into the cell.

### Exocytosis

Exocytosis is the reverse of endocytosis. Material to be secreted from the cell is enclosed in a membrane-bound vesicle that moves towards and fuses with the cell membrane, releasing the material into the extracellular space.

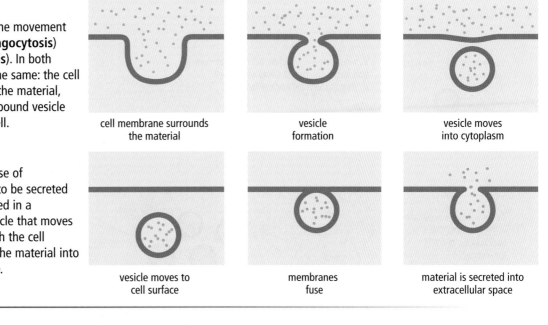

cell membrane surrounds the material

vesicle formation

vesicle moves into cytoplasm

vesicle moves to cell surface

membranes fuse

material is secreted into extracellular space

www.wiley.com/legacy/college/boyer/0470003790/animations/membrane_transport/membrane_transport.htm
www.northland.cc.mn.us/biology/Biology1111/animations/active1.swf

## LET'S THINK ABOUT THIS

The graph opposite shows the effect of oxygen concentration on ion uptake by active transport.

Try to explain:
- why the rate of sodium ion absorption increases and then becomes constant
- the changes to glucose concentration.

Both oxygen and glucose are required to produce ATP in respiration. As the oxygen concentration increases, the rate of respiration increases providing more energy for active transport. The rate of ion absorption therefore increases. However, glucose is being used up in respiration. Eventually, a low glucose concentration limits the rate of ion transport, and the graph levels off.

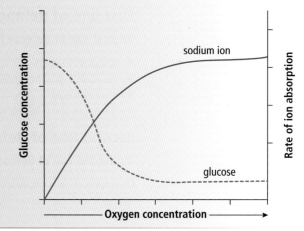

# CELLULAR RESPONSE IN DEFENCE I

Immunity includes all the methods by which the body resists infection by invading microbes.

## VIRUSES

Viruses are very small, ranging in size from 10 to 30 nm. They contain **nucleic acid** (either **DNA or RNA**) surrounded by a **protein coat**.

### Viral replication

Viruses can only replicate by invading another cell and altering that host cell's metabolism to allow replication of viral DNA and RNA, so that new viruses can be produced.

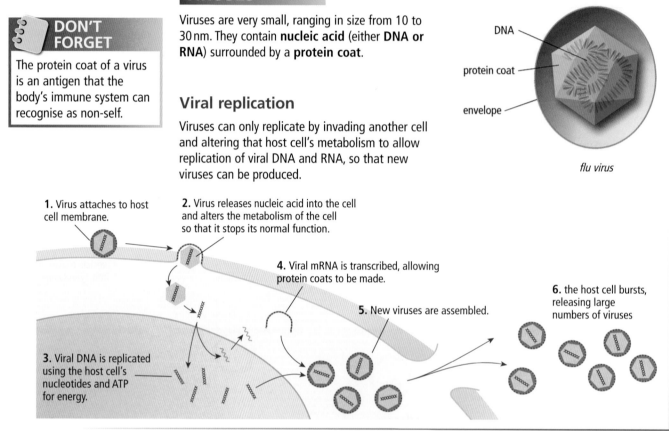

DNA

protein coat

envelope

*flu virus*

**1.** Virus attaches to host cell membrane.

**2.** Virus releases nucleic acid into the cell and alters the metabolism of the cell so that it stops its normal function.

**3.** Viral DNA is replicated using the host cell's nucleotides and ATP for energy.

**4.** Viral mRNA is transcribed, allowing protein coats to be made.

**5.** New viruses are assembled.

**6.** the host cell bursts, releasing large numbers of viruses

> **DON'T FORGET**
>
> The protein coat of a virus is an antigen that the body's immune system can recognise as non-self.

www.whfreeman.com/kuby/content/anm/kb03an01.htm

## INNATE IMMUNITY

Some general body functions that are present at birth (**inborn**) provide first and second lines of defence against invading microbes.

### First line of defence

The **first line of defence** includes:

- enzymes in tears
- resistance of skin epidermis to penetration by microbes
- stomach acid
- mucus in the respiratory tract.

### Second line of defence

The **second line of defence** includes phagocytosis by **macrophages**.

Macrophages are **phagocytes** that circulate around the body removing unwanted cells, antibody–antigen complexes and invading microbes.

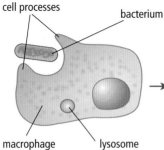
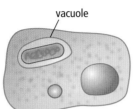

cell processes

bacterium

vacuole

macrophage     lysosome

Cell processes surround the bacterium, enclosing it in a vacuole.

The vacuole moves into the cell.

A lysosome fuses with the vacuole, releasing digestive enzymes.

Enzymes break down the bacterium. Products may be reused by the cell.

## ACQUIRED IMMUNITY

All cells have molecules on their surfaces which act as **antigens**. The body's immune system recognises antigens on its own cells as **self-antigens** and does not attack them. However, the antigens on the membranes of invading microbes or on the cells of transplanted tissue are recognised as being **non-self** or **foreign**, and the immune system tries to destroy them. This is known as an **immune response**.

### The immune response

When a microbe invades, surface antigens are recognised by white blood cells called lymphocytes. Lymphocytes are made in the bone marrow and can be divided into T-lymphocytes and B-lymphocytes.

### T-lymphocytes and the cell-mediated response

T-lymphocytes migrate from the bone marrow to the thymus gland (in the chest cavity) before moving to other lymphatic tissue. These cells do not produce antibodies. There are several types of T-lymphocyte. **Killer T-lymphocytes** recognise foreign antigens that have been left on the surface of infected body cells and act by releasing enzymes that destroy the infected cells. This is called the **cell-mediated response**.

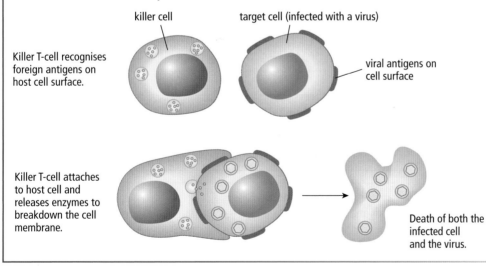

killer cell

target cell (infected with a virus)

Killer T-cell recognises foreign antigens on host cell surface.

viral antigens on cell surface

Killer T-cell attaches to host cell and releases enzymes to breakdown the cell membrane.

Death of both the infected cell and the virus.

## LET'S THINK ABOUT THIS

To produce vaccines, viruses are treated to render them harmless. Which part of the virus must be (**i**) destroyed and (**ii**) retained? Explain.

# CELLULAR RESPONSE IN DEFENCE II

## ACQUIRED IMMUNITY

### B-lymphocytes and the humoral response

**B-lymphocytes** migrate from the bone marrow to tissues of the lymphatic system, such as lymph nodes and spleen. On contact with a foreign antigen, B-lymphocytes divide to form plasma cells that produce antibodies in the **humoral response**. Antibodies are Y-shaped molecules with two receptor sites, one on each arm. The shape of this binding site is **specific** to allow attachment only to the antigen that had initially been recognised. Once the antibody locks on to the antigen, the harmless antigen–antibody complex can be removed by macrophages. B-lymphocytes also give rise to **memory cells**. If the body is invaded again by the same microbe, memory cells mount a rapid response by dividing to produce plasma cells.

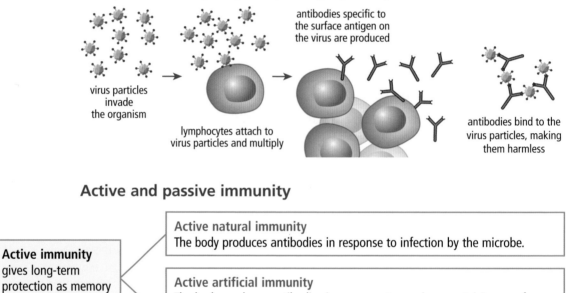

virus particles invade the organism

lymphocytes attach to virus particles and multiply

antibodies specific to the surface antigen on the virus are produced

antibodies bind to the virus particles, making them harmless

### Active and passive immunity

**Active immunity** gives long-term protection as memory cells are produced.

> **Active natural immunity**
> The body produces antibodies in response to infection by the microbe.

> **Active artificial immunity**
> The body produces antibodies in response to vaccines containing a safe form of the microbe. The microbe is made harmless by either chemical or heat treatment that destroys DNA, but keeps surface antigens intact.

**Passive immunity** gives short-term protection as no memory cells are produced.

> **Passive natural immunity**
> Ready-made antibodies are passed from mother to baby either across the placenta or in breast milk. This is important, as the baby cannot make its own antibodies until about 3 months of age.

> **Passive artificial immunity**
> Ready-made antibodies are injected. The antibodies are either collected from another species that has been injected with a safe form of the microbe or may be extracted from human blood plasma.

## ALLERGY

The body sometimes over-reacts and produces an immune response to small traces of a harmless foreign substance. The person is said to have suffered an **allergic response**, and the antigen involved is called an **allergen**. Examples include animal hair and tree pollen. An allergic response involves the production of histamine, which causes inflammation of tissues, dilation of blood vessels and smooth muscle contraction. In the lungs, bronchi become constricted, leading to breathing difficulties. Anaphylaxis, a life-threatening condition, can result when the reaction is acute. Allergies are treated with anti-histamine drugs or, in acute cases, by adrenaline injection.

## AUTO-IMMUNITY

Auto-immune diseases result when the body's immune system mistakes 'self' tissue for 'non-self' and starts to attack. Alterations to 'self' antigens can be caused by drugs, by genetic mutation or by viral action on body cells. In rheumatoid arthritis, the immune system attacks cells of the tendons and joints, causing them to become inflamed and painful. Diabetes mellitus results from a humoral response against the insulin-producing cells in the pancreas.

## LET'S THINK ABOUT THIS

The four main blood groups are categorised by the presence or absence of two types of antigen on the surface of the red blood cells. The table below shows the antigens and antibodies associated with each blood group.

| Blood group | A | B | AB | O |
|---|---|---|---|---|
| Antigens on red blood cells | A antigen | B antigen | A and B antigens | neither A nor B antigens |
| Antibodies in blood plasma | anti-B antibodies | anti-A antibodies | neither anti-A nor anti-B antibodies | both anti-A and anti-B antibodies |

If donor blood containing B antigens is given to patients with group A or group O blood, the antibodies present in their plasma cause the red blood cells to clump together (**agglutination**). Similarly, donor blood containing A antigens agglutinates when mixed with either group B or group AB blood. Blood group O is called the **universal donor** and blood group AB is called the **universal recipient**. Why?

**Primary and secondary response**

You should be able to use the following graph to describe differences between first and subsequent infection by a non-self antigen.

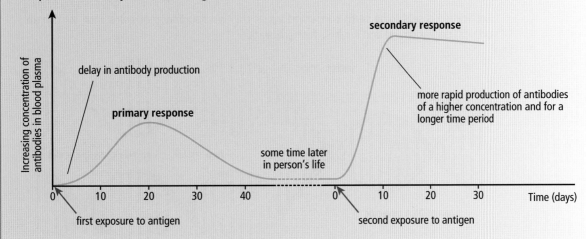

# DNA REPLICATION AND CHROMOSOMES

## DNA REPLICATION

DNA carries the genetic code – the template for protein synthesis (see page 14). When a cell divides, it is important that the resulting daughter cells receive the correct genetic information so that proteins required for normal metabolism can be produced. To allow this to happen, DNA must be copied before either mitosis or meiosis begins. This duplication of DNA is known as **replication**.

In order for replication to occur, the nucleus must contain:

- DNA
- the four types of nucleotide
- ATP
- enzymes.

The process is outlined below.

1  The DNA molecule unwinds.

2  The hydrogen bonds between the bases break and the molecule 'unzips', exposing the bases of both DNA strands.

3  Free DNA nucleotides move in to form complementary base pairs with the exposed bases, and hydrogen bonds form between them.

4  Strong chemical bonds form between the phosphate and deoxyribose sugar of adjacent nucleotides.

5  The new daughter DNA molecules wind up, each forming a double helix.

parent DNA strand

1

hydrogen bonds

2

3

4

DNA nucleotide

Replication is said to be **semi-conservative**, as the two daughter molecules each contain an original parent strand and a newly-formed strand.

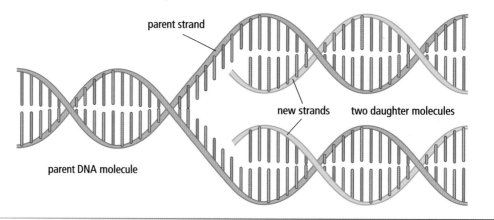

parent strand

new strands    two daughter molecules

parent DNA molecule

# CHROMOSOME COMPLEMENT, KARYOTYPES AND HOMOLOGOUS PAIRS

Each chromosome is composed largely of a DNA molecule with a length ranging between 50 million and 250 million base pairs. The number of chromosomes found in the nuclei of normal body cells is called the **chromosome complement**. Human body cells have a chromosome complement of **46**, arranged as **23 homologous pairs**. One of each pair was inherited from the mother and the other homologous chromosome from the father.

You should remember that the part of a DNA molecule which codes for one protein is called a **gene**. There are thousands of genes on the human chromosomes. Homologous chromosomes carry the same genes (although these may code for different forms of the same characteristic). When arranged side by side, the homologous chromosomes match each other gene for gene, as shown opposite.

To examine the human chromosome complement, a **karyotype** is produced by taking photographs of the chromosomes during cell division and arranging them by shape, size and banding pattern. Because photographs are taken after DNA replication, each chromosome is seen as two DNA strands (called **chromatids**) held together by a **centromere**.

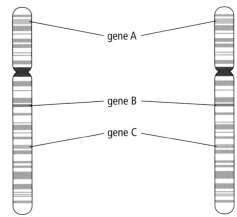

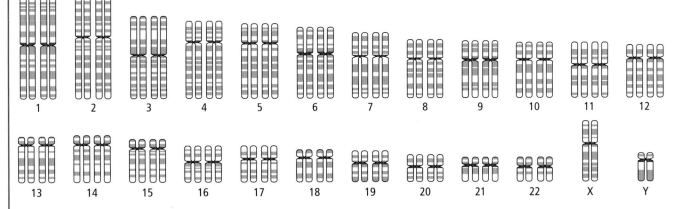

In a karyotype, chromosome pairs 1 to 22 are called **autosomes**. The final pair is the **sex chromosomes** (X and X, or X and Y). Only the sex chromosomes are involved in determining an individual's sex; females carry two X chromosomes and males have one X chromosome and one Y chromosome. Sex chromosomes can be differentiated by their size, the Y chromosome being much smaller than the X chromosome.

**DON'T FORGET**

To remember that males are XY, remember 'Y is he different?'

## LET'S THINK ABOUT THIS

The diagram opposite shows the process of DNA replication.

1 Name components X and Y.

2 Name another substance (not shown in the diagram) that is required for DNA replication.

3 DNA replication is said to be semi-conservative. Explain.

4 Why is DNA replication necessary?

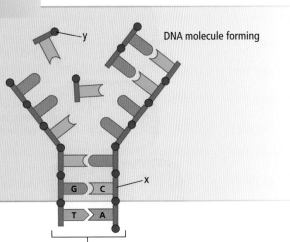

DNA molecule forming

DNA molecule

# MEIOSIS

## MEIOSIS

**Meiosis** is the type of cell division that results in the production of sex cells (**gametes**) and takes place in the testes and the ovaries. The events of meiosis are important for two reasons:

**1** In sexual reproduction, two **gametes** fuse to produce a **zygote**. To maintain the correct number of chromosomes in the new individual, the gametes must contain half the number of chromosomes found in all other cells of the organism. The gametes are said to be **haploid** (contain one set of chromosomes); all other cells are **diploid** (contain two sets of chromosomes).

**2** Meiosis allows genetic **variation** to be introduced. For any species, variation is the key to survival, as the differences it produces may allow some individuals to survive environmental change. Over a period of time, variation can lead to the **evolution** of a new species.

Meiosis consists of two divisions:

- The **first meiotic division (meiosis I)**, which reduces the number of chromosomes.

- The **second meiotic division (meiosis II)**, which reduces the quantity of DNA on each chromosome.

### Meiosis I

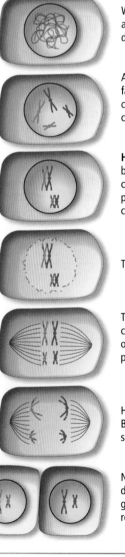

When the cell is not dividing, chromosomes of the gamete mother cell are long, thin and invisible in the cell. Just before the cell begins to divide, the DNA replicates.

As the cell begins to divide, the chromosomes become shorter and fatter (condensed) forming the characteristic X-shape. Each chromosome is composed of two chromatids held together by a centromere.

**Homologous** chromosomes line up together on a gene-for-gene basis. Adjacent chromatids join at several points along their length, called **chiasmata** (singular = chiasma), and exchange segments in a process known as **crossing over**. Once crossing over is complete, the chromatids separate.

The nuclear membrane breaks down.

The spindle invades the central region of the cell, and homologous chromosomes line up together on either side of the equator. The orientation of each chromosome pair is random in relation to all other pairs (**independent assortment**).

Homologous pairs are then pulled apart to opposite poles of the cell. Because the centromere has not been copied, each chromosome is still double-stranded.

Nuclear membranes form, and the cytoplasm divides to produce two daughter cells, each with half the number of chromosomes of the gamete mother cell. But a second meiotic division is now necessary to reduce the number of chromatids on each chromosome.

> **DON'T FORGET**
>
> Variation is introduced during crossing over and independent assortment.

**contd**

## MEIOSIS contd

### Meiosis II

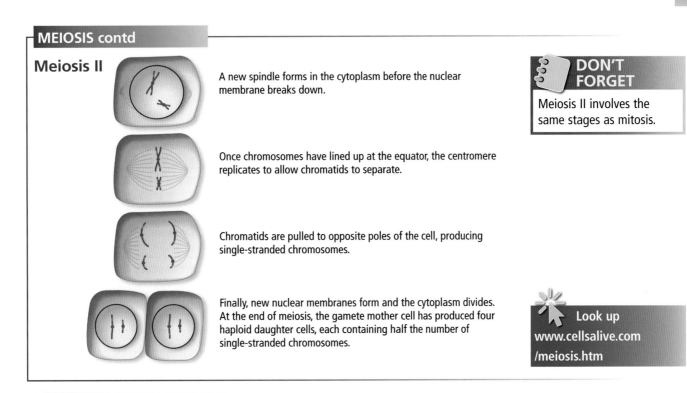

A new spindle forms in the cytoplasm before the nuclear membrane breaks down.

Once chromosomes have lined up at the equator, the centromere replicates to allow chromatids to separate.

Chromatids are pulled to opposite poles of the cell, producing single-stranded chromosomes.

Finally, new nuclear membranes form and the cytoplasm divides. At the end of meiosis, the gamete mother cell has produced four haploid daughter cells, each containing half the number of single-stranded chromosomes.

**DON'T FORGET**

Meiosis II involves the same stages as mitosis.

**Look up**
www.cellsalive.com
/meiosis.htm

## CROSSING OVER

Crossing over takes place in meiosis I. It introduces variation as different combinations of alleles are produced in the gametes.

homologous
chromosomes pair up

chiasmata form
allowing crossing over

recombination of
alleles occurs

## INDEPENDENT ASSORTMENT

During meiosis I, homologous chromosomes line up on either side of the equator. The side of the equator that each chromosome ends up on is entirely random and independent of all the other chromosome pairs. By increasing the possible combinations of chromosomes in any gamete, variation is again increased.

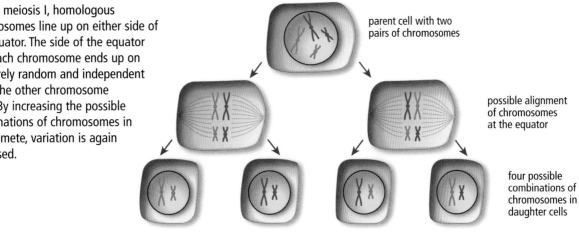

parent cell with two pairs of chromosomes

possible alignment of chromosomes at the equator

four possible combinations of chromosomes in daughter cells

## LET'S THINK ABOUT THIS

Try drawing out all the possible gametes that could result from independent assortment when the gamete mother cell contains a diploid number of six chromosomes. (Hint: there are eight possible combinations of chromosomes.)

gamete mother cell

# MONOHYBRID INHERITANCE

Before you can work through examples of monohybrid inheritance, you must understand and be able to use the following terms.

| Term | Definition |
|------|-----------|
| **Allele** | Alleles are different forms of a gene. For example, the shape of the human ear lobe is determined by a single gene, but there are two alleles – free ear lobe or fixed ear lobe. |
| **Phenotype** | This is the appearance of the characteristic, for example brown hair or blue eyes. |
| **Genotype** | The alleles that are present in an individual, for example AA or Aa. |
| **Homozygous** | Two identical alleles are present in an individual (AA). Because this individual always passes on the same type of allele to its offspring, it is said to be **true-breeding**. |
| **Heterozygous** | Two different alleles are present in an individual (Aa). |
| **Dominant** | A dominant allele, if present in the genotype, is always seen in the phenotype. It is represented in the genotype by a capital letter. |
| **Recessive** | A recessive allele is seen in the phenotype only if no dominant allele is present. It is represented in the genotype by a lower-case letter. |
| $F_1$ | The first generation offspring. |
| $F_2$ | The second generation; offspring from an $F_1$ cross. |

## WORKING THROUGH A MONOHYBRID CROSS

### Tongue rolling

In monohybrid inheritance, a single gene determines the characteristic under consideration. Let's look at the gene that controls tongue rolling. The allele for tongue rolling (T) is dominant to the allele for non-rolling (t). The cross below shows the possible offspring when a person who is homozygous for tongue rolling has children with a non-roller (who is therefore homozygous recessive).

| parent (P) | phenotype | tongue-roller | × | non-roller |
|---|---|---|---|---|
| | genotype | TT | | tt |
| gametes | | All T | | All t |
| $F_1$ | phenotype | | All tongue-rollers | |
| | genotype | | Tt | |

To calculate which genotypes would be present in the second generation, let's do a genetic cross between one of the children and another heterozygous (Tt) individual.

| | tongue-roller $F_1$ | × | tongue-rolling partner |
|---|---|---|---|
| | Tt | | Tt |
| gametes | T or t | | T or t |

**DON'T FORGET**

Make sure that you draw the punnet square on the exam paper so that it is easier to check your answers.

$F_2$ punnet square

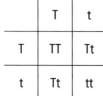

|   | T | t |
|---|---|---|
| **T** | TT | Tt |
| **t** | Tt | tt |

| $F_2$ genotype | $F_2$ phenotype |
|---|---|
| TT or Tt | tongue-roller |
| tt | non-roller |

In this cross, the expected frequency is **3** tongue-rollers : **1** non-roller. However, the actual ratio of dominant to recessive phenotypes may not be 3:1, as **fertilisation is a random process**. As the number of offspring in the population increases, the closer the actual and expected ratios become.

contd

## WORKING THROUGH A MONOHYBRID CROSS contd

## Antigen D

Let's consider the inheritance of **antigen D** in humans. The red blood cells of people who are **Rhesus positive (Rh+)** have this antigen on their surface. There are two alleles:

- The dominant form is the allele for antigen D (let's use an uppercase letter D to represent it).
- The recessive form is the allele which lacks antigen D (use the lowercase letter d to represent it).

This table shows the possible genotypes and phenotypes for antigen D.

A genetic cross between a female who is Rhesus negative (dd) and a man who is Rhesus positive (DD) would look like this:

| Genotype | Phenotype |
|----------|-----------|
| DD or Dd | Rhesus positive |
| dd | Rhesus negative |

| | | | | |
|---|---|---|---|---|
| parents (P) | phenotype | Rhesus negative female | × | Rhesus positive male |
| | genotype | dd | | DD |
| gametes | | all d | | all D |
| | | | | |
| $F_1$ | genotype | Dd | | |
| | phenotype | Rhesus positive | | |

## Cystic fibrosis

People who suffer from **cystic fibrosis** produce a very sticky form of mucus which is difficult to remove from the respiratory passages in the lungs. The disease is caused by an **autosomal recessive** allele. Consider an example where a woman who suffers from cystic fibrosis has a child with a man who is heterozygous for this gene. In this cross, the recessive allele which causes cystic fibrosis is represented by the letter r and the normal allele is represented by the letter R.

| | | | | |
|---|---|---|---|---|
| parents (P) | | mother | × | father |
| | phenotype | cystic fibrosis | | unaffected |
| | genotype | rr | | Rr |
| gametes | | all r | | R or r |

$F_1$ punnet square

| | R | r |
|---|---|---|
| r | Rr | rr |
| r | Rr | rr |

| F₁ offspring | | |
|----------|-----------|----------|
| Genotype | Phenotype | Expected frequency |
| Rr | unaffected | 50% |
| rr | affected | 50% |

## LET'S THINK ABOUT THIS

Using tongue-rolling as your example, calculate the expected frequency of each phenotype in the first generation of the following crosses.

| | Mother's genotype | Father's genotype |
|---|-------------------|-------------------|
| 1 | TT | Tt |
| 2 | tt | TT |
| 3 | Tt | Tt |

# MULTIPLE ALLELES AND LEVELS OF DOMINANCE

## MULTIPLE ALLELES

**DON'T FORGET**

Each individual can inherit only two of the alleles.

If a gene has more than two alleles, it is said to have **multiple alleles**.

The gene that codes for the ABO blood-group system (see page 27) has three alleles: A, B, and O. The table opposite shows the possible genotypes and phenotypes.

| Genotype | Phenotype |
|----------|-----------|
| AA or AO | A |
| BB or BO | B |
| OO | O |
| AB | AB |

## CODOMINANCE

In codominance, two alleles are equally dominant, neither allele being recessive to the other. As a result, both alleles are expressed in the phenotype of heterozygous individuals. Both the **ABO** and **MN** blood-group systems are examples of codominance.

### ABO blood-group system

Of the three alleles involved (A, B and O), A and B are codominant and O is recessive to both A and B. The following example looks at the children that could arise in a family where the mother is heterozygous for blood group A and the father is heterozygous for blood group B.

| parents | | mother | | father |
|---------|--|--------|--|--------|
| phenotype | | A | × | B |
| genotype | | AO | | BO |
| gametes | | A or O | | B or O |

$F_1$ punnet square

| | A | O |
|---|---|---|
| B | AB | BO |
| O | AO | OO |

| Genotype | Phenotype | Expected frequency |
|----------|-----------|--------------------|
| AO | A | 25% |
| BO | B | 25% |
| OO | O | 25% |
| AB | AB | 25% |

### MN blood-group system

Antigens M and N are also found on the surface of red blood cells. The gene involved has two alleles (M and N) which are codominant. Possible blood groups are shown in the table opposite.

Consider a family where the parents both have blood group MN.

| Genotype | Phenotype (Blood group) | Antigens present on red blood cells |
|----------|-------------------------|-------------------------------------|
| MM | M | M |
| NN | N | N |
| MN | MN | M and N |

**DON'T FORGET**

When alleles are codominant, two different letters (both capitals) are used in the genotype.

| parents | | mother | | father |
|---------|--|--------|--|--------|
| phenotype | | MN | × | MN |
| genotype | | MN | | MN |
| gametes | | M or N | | M or N |

contd

## CODOMINANCE contd

F$_1$ punnet square

|   | M | N |
|---|---|---|
| M | MM | MN |
| N | MN | NN |

| F$_1$ offspring | | |
|---|---|---|
| Genotype | Phenotype (blood group) | Expected Frequency |
| MM | M | 25% |
| NN | N | 25% |
| MN | MN | 50% |

## INCOMPLETE DOMINANCE

Where two alleles are **incompletely dominant**, a heterozygous individual will have a phenotype that is a blend of the two homozygous types. An example of incomplete dominance is sickle-cell trait. The gene which codes for haemoglobin has two alleles: allele A codes for normal haemoglobin, and allele S codes for haemoglobin S. Unaffected individuals have a genotype AA, and their red blood cells contain normal haemoglobin. The table below shows the other genotypes and phenotypes associated with this gene.

| Genotype | Phenotype | Description of condition |
|---|---|---|
| SS | sickle-cell anaemia | Individuals who are homozygous for the haemoglobin S allele suffer from sickle-cell anaemia. Haemoglobin S binds with less oxygen than normal haemoglobin. Red blood cells are sickle-shaped and stick together, causing clumping which blocks blood vessels. People with this condition often die at a young age. |
| AS | sickle-cell trait | Heterozygous individuals suffer a less serious condition called sickle-cell trait. Red blood cells are a normal biconcave shape and contain both normal haemoglobin and haemoglobin S. |

In the following example, a woman who has sickle-cell trait and a man who is unaffected have children together. The expected frequency of each phenotype in the first generation is shown.

| parents | mother | | father |
|---|---|---|---|
| phenotype | sickle-cell trait | × | unaffected |
| genotype | AS | | AA |
| gametes | A or S | | all A |

F$_1$ punnet square

|   | A | A |
|---|---|---|
| A | AA | AA |
| S | AS | AS |

| Genotype | Phenotype | Expected frequency |
|---|---|---|
| AA | unaffected | 50% |
| AS | sickle-cell trait | 50% |

## LET'S THINK ABOUT THIS

Work through the following genetic crosses, calculating the percentage chance for each possible F$_1$ phenotype.

|   | Mother's phenotype | Father's phenotype |
|---|---|---|
| 1 | blood group M | blood group N |
| 2 | blood group AB | blood group AB |
| 3 | sickle-cell anaemia | unaffected |
| 4 | sickle-cell trait | sickle-cell trait |

# POLYGENIC AND SEX-LINKED INHERITANCE

## POLYGENIC INHERITANCE

Sometimes characteristics are determined by two or more genes working together. This pattern of inheritance is called **polygenic inheritance.** Polygenic characteristics display **continuous variation** – a wide range of phenotypes are present in a **normal distribution** (see below). In humans, **eye colour**, **height** and **skin colour** are examples of characteristics with polygenic inheritance.

### Inheritance of skin colour

| Skin colour | Genotype | Number of dominant alleles present |
|---|---|---|
| white | aabb | 0 |
| light | Aabb or aaBb | 1 |
| medium | AAbb, aaBB or AaBb | 2 |
| dark | AABb or AaBB | 3 |
| very dark | AABB | 4 |

Although many genes are actually involved in the inheritance of skin colour, we will consider a simplified form of inheritance in which only two genes (A and B) control the characteristic. The dominant allele for each gene codes for the production of melanin, the pigment in the skin. The recessive allele does not code for melanin production. The effect of the dominant gene is cumulative: that is, the more dominant alleles present in the genotype, the darker the skin colour. The table shows how the number of dominant alleles is related to skin colour.

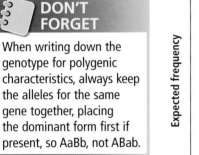

**DON'T FORGET**

When writing down the genotype for polygenic characteristics, always keep the alleles for the same gene together, placing the dominant form first if present, so AaBb, not ABab.

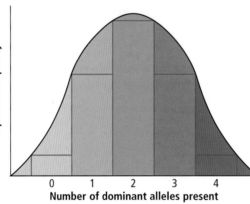

**Number of dominant alleles present**

The graph showing the frequency of each skin colour in the population fits under a bell-shaped curve (a **normal distribution**), with the majority of people inheriting an intermediate skin colour.

## SEX-LINKED INHERITANCE

Although the X and Y chromosomes are homologous chromosomes, the difference in their size means that some genes are found on both chromosomes and others are found only on the larger X chromosome.

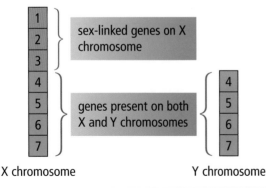

sex-linked genes on X chromosome

genes present on both X and Y chromosomes

X chromosome          Y chromosome

In the diagram, genes 1–3 are found only on the X chromosome and are called **sex-linked genes.** A male inherits one X and one Y chromosome – and so, for the sex-linked genes, he will only have one allele (on the X chromosome). This means that the characteristic of the gene on the X chromosome will be expressed in a male, no matter if a dominant or recessive allele has been inherited (remember that, in non-sex-linked genes, two recessive alleles must be present to display the recessive phenotype).

contd

## SEX-LINKED INHERITANCE contd

### Red–green colour blindness in humans

Normal vision (N) is dominant and red–green colour blindness (n) is recessive. This gene is a sex-linked gene, which means it is located on the X chromosome.

Consider a cross between a female homozygous for normal vision and a colour-blind male.

| parents | $X^N X^N$ | × | $X^n Y$ |
|---|---|---|---|
| gametes | all $X^N$ | | $X^n$ or Y |
| $F_1$ | $X^N X^n$ | or | $X^N Y$ |
| | carrier female | | normal male |

The $F_1$ female has inherited the colour-blindness gene from her father but, as she has also inherited a dominant gene for normal sight from her mother, she has normal vision. She is, however, a **carrier** and could pass on the colour-blindness gene to her offspring.

Red–green colour blindness is quite rare in females because they have two X chromosomes and so would have to inherit two recessive alleles. In males, inheriting an X chromosome with a recessive gene will give rise to colour blindness.

The inheritance of colour blindness can be traced through generations using a family tree.

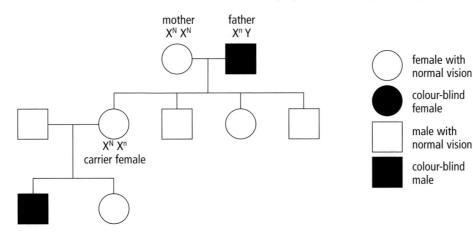

This family tree shows that sex-linked characteristics can skip generations. The male parent is affected but none of his children is colour blind. One of his daughters is a carrier, and she has passed on the colour-blindness gene to her son. More males than females are affected.

> **DON'T FORGET**
>
> Sex-linked genes are always indicated using the letters X and Y in the genotype.

> **DON'T FORGET**
>
> If representing genes with letters that are very similar in upper and lower case, for example C and c, make sure they are clearly different sizes so that examiners can see which allele you are referring to.

### LET'S THINK ABOUT THIS

You should be able to work out the theoretical outcomes of crosses involving sex-linked genes.

**1** What are the possible $F_1$ genotypes and phenotypes from the following crosses?

| | Female | | Male | |
|---|---|---|---|---|
| | genotype | phenotype | genotype | phenotype |
| Cross 1 | $X^N X^n$ | carrier | $X^n Y$ | colour blind |
| Cross 2 | $X^N X^n$ | carrier | $X^N Y$ | normal vision |

**2** Haemophilia is another sex-linked condition. Write down all the possible genotypes and phenotypes associated with this gene.

# MUTATIONS I

## MUTAGENIC AGENTS

A **mutation** is an alteration to the genetic information in a cell. **Random** mutations occur naturally within a population, usually at a very **low frequency**, and introduce variation into the species. However, some agents cause the rate of mutation to increase. Examples of **mutagenic agents** include irradiation by ultra-violet light, gamma rays, or X-rays; and a variety of chemicals such as those in mustard gas and cigarette smoke.

## GENE MUTATIONS

A gene is a segment of a DNA molecule that carries the code for one protein. The code takes the form of a sequence of bases, with a triplet of bases coding for one amino acid (see page 14). If the sequence of bases is altered, the corresponding sequence of amino acids may change, possibly altering the protein produced. There are four types of gene mutation:

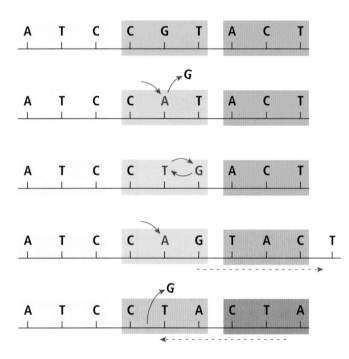

normal strand of DNA

**Substitution** – one of the bases is replaced by a different base (here A replaces G). This is a **point mutation**, as it only changes the amino acid coded for by the affected triplet.

**Inversion** – some of the bases swap position (here G, T has been inverted to give T, G). This is another **point mutation**, as only the amino acid coded for by the affected triplet is altered.

**Insertion** – an extra base (A) has been inserted into the sequence, moving the bases after the insertion one place to the right. This is a **frame-shift mutation**, as every amino acid after the point of insertion is altered.

**Deletion** – a base has been removed (G) from the sequence, shifting the bases after the deletion to the left. This is another **frame-shift mutation**, as every amino acid after the point of deletion is altered.

Look up http://www.bbc.co.uk/scotland/learning/bitesize/higher/biology/genetics_adaptation/mutations_rev1.shtml

As point mutations usually alter just one amino acid, the effect on any protein produced is generally minor. However, sometimes the amino acid that is altered is in a key position for protein function. An example of this is sickle-cell anaemia, where normal haemoglobin is replaced by haemoglobin S.

Frame-shift mutations always cause major changes to the amino acid sequence and therefore greatly alter the protein that is synthesised. Frame-shift mutations frequently result in non-viable gametes.

## CHANGES TO CHROMOSOME NUMBER

During cell division, the spindle moves chromosomes from the equator of the cell to the poles. Sometimes the spindle doesn't function properly, and chromosomes are not pulled apart, ending up in the wrong cell. This is called **non-disjunction** and results in the production of gametes that have either extra or missing chromosomes. The diagram below shows non-disjunction in meiosis I.

| homologous chromosomes line up at equator | the spindle fails to separate one of the chromosome pairs | one daughter cell has an extra chromosome and the other is missing one |

Non-disjunction can affect non-sex chromosomes (**autosomes**) or the sex chromosomes.

### Non-disjunction of autosomes

Non-disjunction of chromosome number **21** causes **Down's Syndrome**. Here, the individual has an extra chromosome number 21 and has a characteristic appearance including prominent, slanted eyelids and a short nose. Mental disability also results. The karyotype below shows the chromosome complement of a male individual with Down's Syndrome.

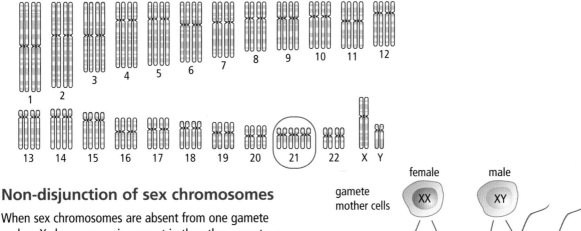

### Non-disjunction of sex chromosomes

When sex chromosomes are absent from one gamete and an X chromosome is present in the other gamete, a condition known as **Turner's Syndrome** results. Individuals are infertile females (XO).

**Kleinfelter's Syndrome** is also caused by non-disjunction of sex chromosomes. This time, an egg containing two X chromosomes is fertilised by a sperm containing a Y chromosome to produce an infertile male (XXY).

## LET'S THINK ABOUT THIS

The incidence of Down's Syndrome increases with maternal age. This increase in frequency is related to the length of time that the spindle is in position, because the gamete mother cells in the ovary enter metaphase I when the woman is still a foetus herself. Meiosis I does not continue until some point after puberty, when that particular gamete mother cell is selected for further development in the ovarian cycle. It may be several decades after forming that the spindle pulls on the chromosomes to separate them.

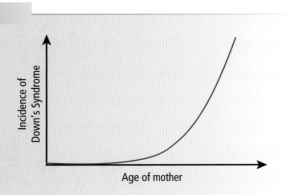

# MUTATIONS II

## GENETIC SCREENING AND COUNSELLING

If a couple may be at risk of having a child with a genetic disorder, they can consult a **genetic counsellor**. The counsellor can calculate the probability of having a child with a genetic disorder and can discuss the potential consequences for the child and family.

The genetic counsellor may construct a **family tree (pedigree)** to map out the history of the disease in the family. Counselling involves discussion of the prenatal and postnatal tests that are available and the different courses of action that may be taken if the tests prove positive. Counselling is important, as it helps couples decide if they should:

**1** have children
**2** use prenatal screening and consider termination of affected foetuses
**3** use *in vitro* fertilisation in conjunction with genetic screening.

> **DON'T FORGET**
>
> Chorionic villus sampling can be carried out earlier than amniocentesis but has a higher risk of miscarriage.

## PRENATAL TESTING

There are two methods of prenatal screening:

**1 amniocentesis**     **2 chorionic villus sampling**.

### Amniocentesis

A needle is inserted through the walls of the abdomen and the uterus into the amniotic sac, to collect a sample of amniotic fluid. Amniotic fluid contains skin and hair cells from the foetus. The extracted cells are then grown in tissue culture before their chromosomes are examined in a karyotype.

Amniocentesis has a low risk of miscarriage (0·5–1% above normal) but cannot be performed until between 16 and 18 weeks of pregnancy, when the mother is beginning to feel an attachment to her unborn child.

### Chorionic villus sampling

A small sample of foetal cells from the placenta is removed either through the wall of the abdomen or through the vagina. The extracted cells are grown in tissue culture and their chromosomes are examined in a karyotype.

This test can be carried out at between 8 to 10 weeks, giving it an advantage over amniocentesis. However, as part of the placenta is being removed, the risk of miscarriage is increased (2–3% above normal) and some maternal cells may be extracted with the foetal cells, giving a false result.

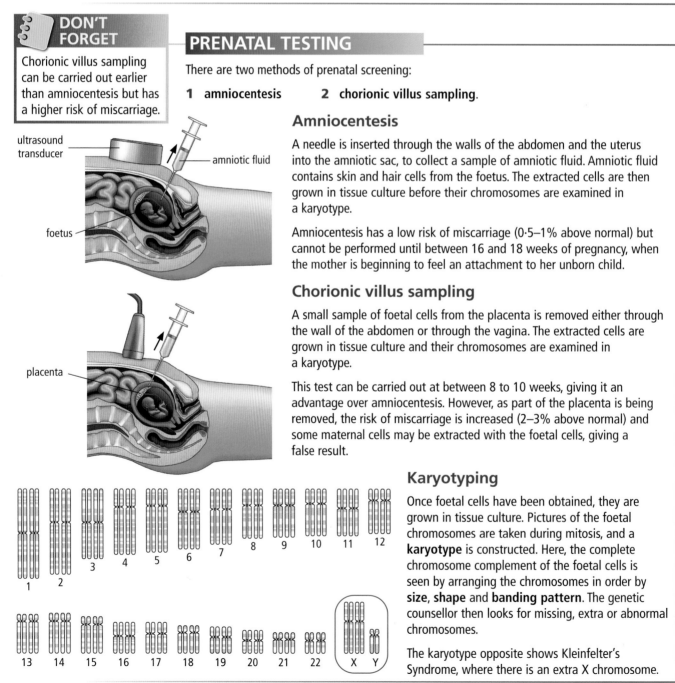

ultrasound transducer
amniotic fluid
foetus
placenta

1  2  3  4  5  6  7  8  9  10  11  12
13  14  15  16  17  18  19  20  21  22  X  Y

### Karyotyping

Once foetal cells have been obtained, they are grown in tissue culture. Pictures of the foetal chromosomes are taken during mitosis, and a **karyotype** is constructed. Here, the complete chromosome complement of the foetal cells is seen by arranging the chromosomes in order by **size**, **shape** and **banding pattern**. The genetic counsellor then looks for missing, extra or abnormal chromosomes.

The karyotype opposite shows Kleinfelter's Syndrome, where there is an extra X chromosome.

## POSTNATAL TESTING

After birth, it is possible to carry out biochemical tests for several inherited conditions. The example that you must be familiar with is PKU (see page 9). Routine screening for PKU is currently undertaken in the UK after every birth. Affected individuals can manage their condition with a reduced-phenylalanine diet.

## FAMILY TREES

### Autosomal dominant inheritance

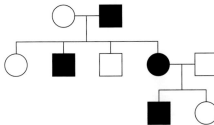

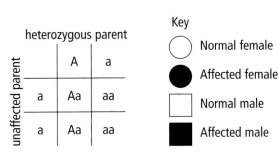

**Key**
- ○ Normal female
- ● Affected female
- □ Normal male
- ■ Affected male

- Parents who are both unaffected have unaffected children.
- Affected individuals with unaffected partners have a 50% chance of having an affected child (see punnet square above).
- The characteristic is seen in every generation.
- Both sexes are equally affected.

Examples of autosomal dominant conditions are **Huntington's chorea** and **achondroplasia**.

### Autosomal recessive inheritance

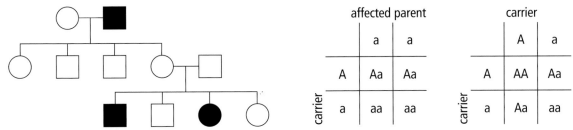

- Parents who are both heterozygous can have affected children.
- Children of an affected parent always inherit the recessive allele.
- Both sexes are equally affected.
- The condition can skip generations.

Examples of autosomal recessive conditions are **cystic fibrosis** and **PKU**.

### Sex-linked inheritance

- More males than females are affected.
- Only females can be carriers.
- Daughters of affected males are either carriers or affected themselves.
- A carrier with a normal partner has a 50% chance of having an affected son or a carrier daughter.
- The condition can skip generations.

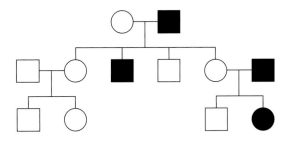

Examples of sex-linked inheritance are **Duchenne muscular dystrophy**, **haemophilia** and **red–green colour blindness**.

## LET'S THINK ABOUT THIS

You should be aware that it is more difficult to assess accurately the probability of inheriting a genetic disorder when the condition is polygenic. This is because environmental factors, as well as genes, play a role in determining the severity of the condition.

# MALE REPRODUCTIVE SYSTEM

The male produces **gametes** (**sperm cells**) continuously from puberty until death and is said to display **continuous fertility**. This is possible because male sex hormone levels remain constant after puberty.

## ORGANS OF THE MALE REPRODUCTIVE SYSTEM

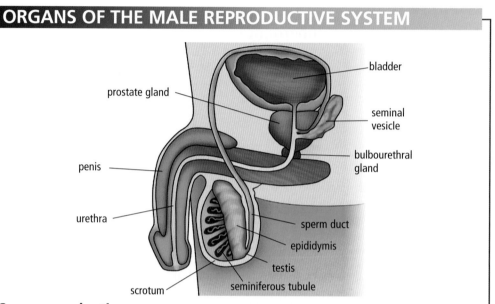

## Sperm production

Sperm cells are produced in the **seminiferous tubules** inside the **testes**. These coiled tubes are surrounded by blood vessels and clusters of **interstitial cells** that produce the hormone **testosterone**. Meiosis of gamete mother cells in the wall of the seminiferous tubules gives rise to immature sperm cells which are released by supporting cells into the lumen of the seminiferous tubule.

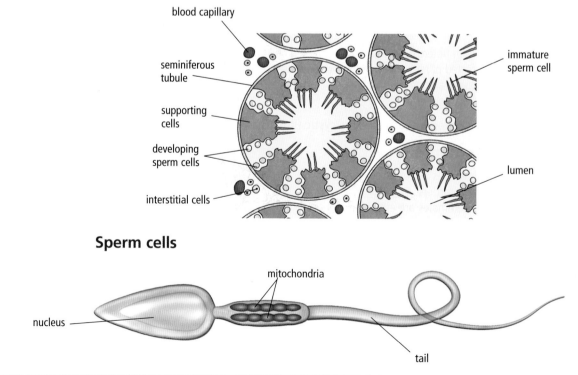

## Sperm cells

## SEMEN

During ejaculation, **semen** passes through the sperm duct and the urethra to be ejected into the female reproductive tract. Semen is a mixture of sperm and secretions from the **seminal vesicles** and **prostate gland**.

### Prostate gland

The prostate gland secretes a milky fluid containing **enzymes** that maintain the thin consistency of the fluid.

### Seminal vesicles

The seminal vesicles secrete an alkaline, viscous fluid containing **fructose** and **prostaglandins**. Fructose is a respiratory substrate that provides energy for movement of the sperms' tails. Prostaglandins cause the female reproductive tract to contract, helping sperm movement.

**DON'T FORGET**

A large number of sperm are needed, as only a tiny fraction reach the ovum.

## HORMONAL CONTROL OF THE MALE REPRODUCTIVE SYSTEM

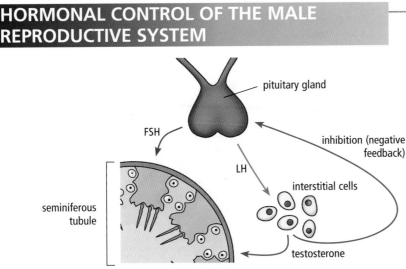

Sperm production is under the control of:

- two hormones released from the pituitary gland in the brain, FSH and LH
- testosterone released from the testes.

### Follicle-stimulating hormone (FSH)

FSH acts on the seminiferous tubules to promote sperm production.

### Luteinising hormone (LH)

LH acts on the interstitial cells of the testes, stimulating the production of **testosterone**.

### Testosterone

Testosterone is produced by the interstitial cells in the testes. It acts on the seminiferous tubules, stimulating sperm production.

Over-production of testosterone is prevented by a **negative feedback** mechanism. When the testosterone level increases above normal, it inhibits secretion of FSH and LH from the pituitary. Testosterone production then stops until the level drops below normal, when the inhibitory effect is switched off and production begins again.

**DON'T FORGET**

Hormones reach their target organs through the bloodstream.

**LET'S THINK ABOUT THIS**

Sterilisation in the male (vasectomy) involves cutting the sperm duct within the scrotum. What effect would this operation have on (**i**) the levels of testosterone in the blood and (**ii**) sperm production?

# FEMALE REPRODUCTIVE SYSTEM

The female displays **cyclical fertility** (eggs are released only once a month) due to changes in hormone levels throughout the menstrual cycle. This continues from puberty to the menopause.

## MENSTRUAL CYCLE

The menstrual cycle consists of cyclical changes in both the ovary and the uterus.

### Changes in the ovary

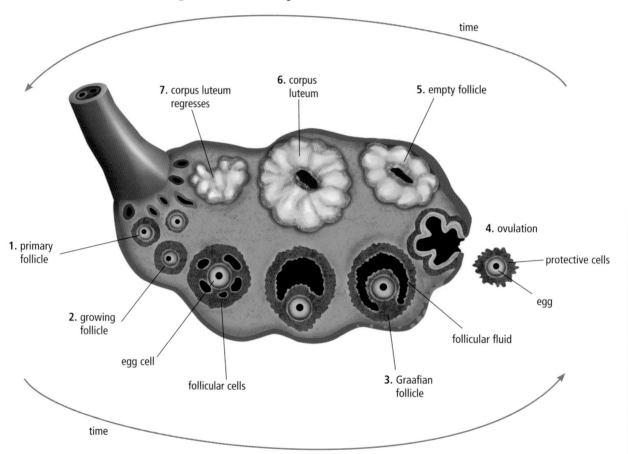

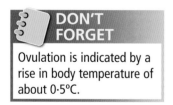

**DON'T FORGET**

Ovulation is indicated by a rise in body temperature of about 0·5°C.

Under the influence of **follicle-stimulating hormone (FSH)** from the pituitary gland, **follicles** begin to develop. The follicular cells produce follicular fluid, which gathers within the follicle as it enlarges. They also secrete **oestrogen**, which acts on the pituitary gland, stimulating the production of **luteinising hormone (LH)**. Usually, only one follicle will mature fully to produce a **Graafian follicle.**

About 10–12 hours after an **LH surge**, usually on about day 14 of the menstrual cycle, the Graafian follicle ruptures, releasing an ovum (**ovulation**). LH now stimulates the development of the **corpus luteum** from the remaining cells of the ruptured follicle. The corpus luteum produces **progesterone** and **oestrogen**, which have a negative feedback effect on the pituitary gland, inhibiting the release of FSH and preventing development of any more follicles.

If no pregnancy occurs, progesterone and oestrogen levels decrease and the corpus luteum degenerates, allowing the development of follicles to begin again in the next cycle. If there is a pregnancy, the corpus luteum continues to function until the placenta is large enough to take over hormone production.

contd

## MENSTRUAL CYCLE contd

### Changes in the uterus

Changes in the uterus occur in three stages during the menstrual cycle:

> **Menstrual stage**
> Low levels of oestrogen and progesterone cause the layer lining the uterus (endometrium) to be shed. The menstrual flow consists of a mixture of endometrial cells, mucus, blood and tissue fluid.

> **Proliferative stage**
> Increasing oestrogen production from the ovarian follicles stimulates repair of the endometrium. The endometrium becomes thicker and develops a good blood supply.

> **Secretory phase**
> **After ovulation**, the endometrium continues to thicken under the influence of progesterone from the corpus luteum. This is accompanied by further development of the endometrial blood vessels and endometrial glands, preparing the uterus for implantation should fertilisation take place.
>
> If there is no fertilisation, the levels of oestrogen and progesterone decrease and the uterus enters the next menstrual phase. If fertilisation does occur, the endometrium will be maintained.

**DON'T FORGET**

Before ovulation, the cervical mucus is thin and watery. After ovulation, the viscosity of the mucus increases.

http://sumanasinc.com/webcontent/animations/content/ovarianuterine.html

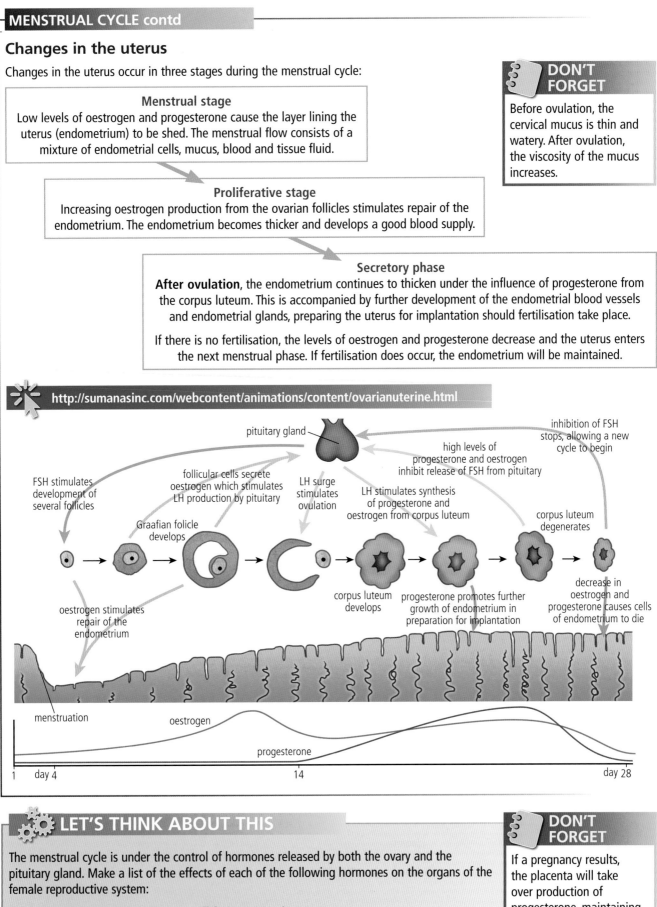

**LET'S THINK ABOUT THIS**

The menstrual cycle is under the control of hormones released by both the ovary and the pituitary gland. Make a list of the effects of each of the following hormones on the organs of the female reproductive system:

**(a)** follicle-stimulating hormone
**(b)** luteinising hormone
**(c)** oestrogen
**(d)** progesterone.

**DON'T FORGET**

If a pregnancy results, the placenta will take over production of progesterone, maintaining the endometrium and preventing miscarriage.

# PRENATAL DEVELOPMENT

If gametes are viable, the following sequence of events can occur after intercourse.

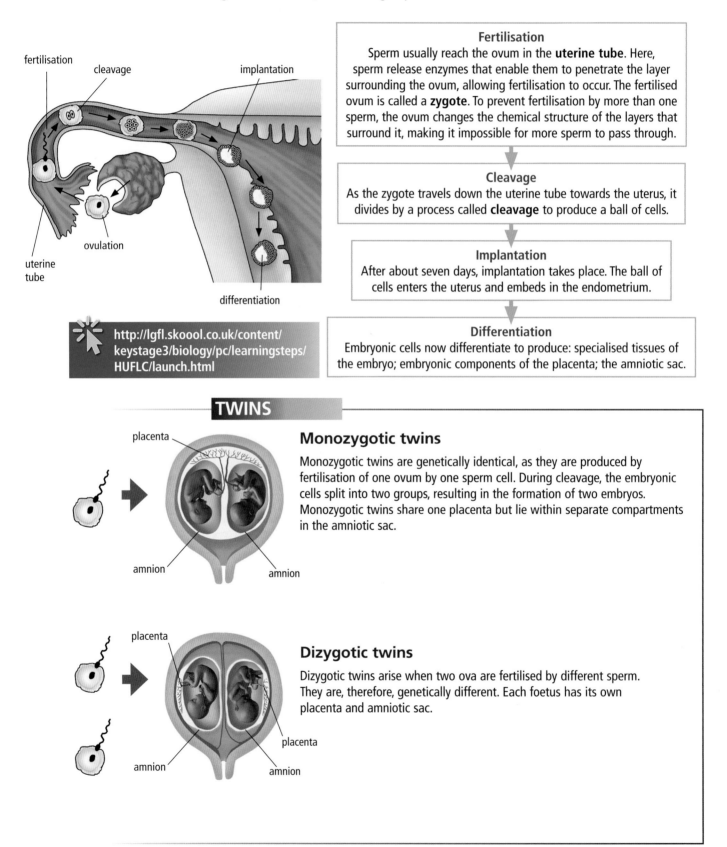

### Fertilisation
Sperm usually reach the ovum in the **uterine tube**. Here, sperm release enzymes that enable them to penetrate the layer surrounding the ovum, allowing fertilisation to occur. The fertilised ovum is called a **zygote**. To prevent fertilisation by more than one sperm, the ovum changes the chemical structure of the layers that surround it, making it impossible for more sperm to pass through.

### Cleavage
As the zygote travels down the uterine tube towards the uterus, it divides by a process called **cleavage** to produce a ball of cells.

### Implantation
After about seven days, implantation takes place. The ball of cells enters the uterus and embeds in the endometrium.

### Differentiation
Embryonic cells now differentiate to produce: specialised tissues of the embryo; embryonic components of the placenta; the amniotic sac.

http://lgfl.skoool.co.uk/content/
keystage3/biology/pc/learningsteps/
HUFLC/launch.html

## TWINS

### Monozygotic twins

Monozygotic twins are genetically identical, as they are produced by fertilisation of one ovum by one sperm cell. During cleavage, the embryonic cells split into two groups, resulting in the formation of two embryos. Monozygotic twins share one placenta but lie within separate compartments in the amniotic sac.

### Dizygotic twins

Dizygotic twins arise when two ova are fertilised by different sperm. They are, therefore, genetically different. Each foetus has its own placenta and amniotic sac.

## PLACENTA

The placenta is the site of exchange of materials between maternal and foetal blood. Within the placenta, foetal blood vessels lie inside finger-like projections called villi. Villi provide a large surface area for exchange of materials with the maternal blood that surrounds them.

maternal blood vessel

villi bathed in maternal blood

intervillous spaces filled with maternal blood

umbilical cord

foetal blood vessels

### Functions of the placenta

**Exchange of materials**

| Substances passing from maternal to foetal blood | | Substances passing from foetal to maternal blood | |
|---|---|---|---|
| **Substance** | **Method of transport** | **Substance** | **Method of transport** |
| oxygen | diffusion | carbon dioxide | diffusion |
| glucose | active transport | urea | active transport |
| antibodies | pinocytosis | | |

**Hormone production**

The placenta takes over production of oestrogen and progesterone from the corpus luteum. These hormones maintain the lining of the uterus throughout pregnancy and prepare the tissues of the mammary glands for milk production.

**Placental barrier**

Although the placenta acts as a protective barrier, some chemicals and microbes can pass from the maternal to the foetal circulation. These include:
- microbes such as the rubella virus
- chemicals such as alcohol, heroin, thalidomide and nicotine.

**DON'T FORGET**

As their blood groups may be incompatible, it is important that maternal blood and foetal blood never mix.

## RHESUS FACTOR

The placenta normally acts as an immune barrier, preventing the mother's body from recognising foetal antigens as foreign. However, should maternal and foetal blood mix during either miscarriage or birth, the mother's immune system will respond by producing antibodies. If the blood of a Rhesus negative (Rh⁻) mother mixes with blood from a Rhesus positive (Rh⁺) baby during birth, the mother's immune system produces anti-D antibodies and memory cells. During subsequent pregnancies, an Rh⁺ foetus will be attacked by anti-D antibodies. To prevent this immune response, a blood transfusion of Rh⁻ blood can be given in subsequent pregnancies to Rh⁺ babies, or the Rh⁻ mother can be given an injection of anti-D antibodies after the birth of the first baby, destroying any of the baby's Rh⁺ blood cells in the mother's blood.

**DON'T FORGET**

Antibodies cross the placenta from mother to foetus.

**DON'T FORGET**

Make sure that you review the Rhesus genetic crosses on page 33.

## BIRTH

During pregnancy, uterine contractions are prevented by high levels of **progesterone**. Towards the end of gestation, progesterone levels decrease and this inhibition ceases. Under stimulation by **oestrogen**, the **pituitary gland** releases **oxytocin**, a hormone that causes the smooth muscle of the uterine wall to contract rhythmically, inducing **labour**. Oxytocin continues to stimulate uterine contractions for a short time after birth, expelling the placenta, preventing excessive bleeding and reducing the size of the uterus.

Gestation lasts approximately 40 weeks, with the expected date of arrival being 280 days after the last menstrual period.

## LET'S THINK ABOUT THIS

Labour can be induced artificially. How?

# LACTATION, CONTRACEPTION AND INFERTILITY

## LACTATION

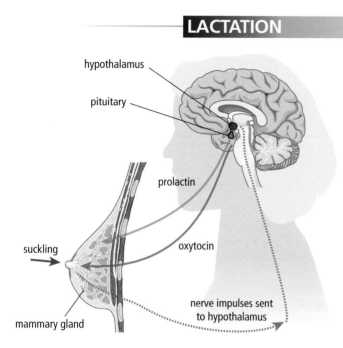

hypothalamus

pituitary

prolactin

oxytocin

suckling

mammary gland

nerve impulses sent to hypothalamus

Milk production is stimulated by the hormone **prolactin**, produced by the pituitary gland. Before birth, progesterone inhibits milk production. However, delivery of the placenta during labour means that this inhibition ends.

During **suckling**, nerve impulses pass from the nipple to the **hypothalamus** in the brain. The hypothalamus stimulates release of **prolactin** and **oxytocin** from the pituitary gland. Oxytocin induces contraction of smooth muscle in the mammary glands, resulting in **milk ejection**. By producing prolactin at the same time, the mammary glands are stimulated to produce more milk for the next feed.

The first liquid produced by the mammary glands is called **colostrum**. Colostrum contains very little lactose and no fat, but is rich in protein and contains **maternal antibodies** (important as the baby cannot make its own antibodies until about 3 months of age). True milk starts to be produced a few days after birth and has a higher concentration of lactose and fat, but fewer antibodies than colostrum.

## BIOLOGICAL BASIS OF CONTRACEPTION

**DON'T FORGET**

The most successful method of contraception is the contraceptive pill.

Contraceptive methods act by preventing either fertilisation or implantation. You should be familiar with the use of hormonal methods and determination of the fertile period, although there are also other methods.

### Hormonal methods

The contraceptive pill contains a combination of the hormones oestrogen and progesterone. These inhibit production of FSH and LH from the pituitary gland. As a result, follicles in the ovary do not mature and ovulation does not take place.

### Determination of the fertile period

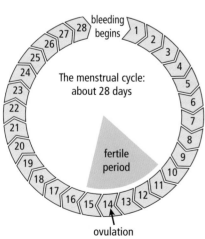

bleeding begins

The menstrual cycle: about 28 days

fertile period

ovulation

Females may try to prevent conception by avoiding intercourse around the fertile period, when conception is possible. However, it should be remembered that this method of contraception is unreliable. As ova survive for approximately 24 hours after ovulation and sperm can live for about 3 days in the female reproductive tract, the fertile period starts about 3 days before ovulation and ends the day after ovulation.

In the example shown, the menstrual cycle lasts 28 days, with ovulation occurring on day 14. The fertile period starts on day 11, as sperm deposited in the female reproductive tract on this day would still be capable of fertilising an ovum on day 14. Day 15 is the last day of the fertile period, as the ovum would not be viable after this day.

Both body temperature and the consistency of cervical mucus are indicators of the fertile period:

● as ovulation approaches, the cervical mucus becomes thinner
● body temperature increases by about 0·5°C after ovulation.

www.nhs.uk/Conditions/Contraception/Pages/How-does-it-work.aspx

## INFERTILITY

For fertilisation and implantation to occur, viable gametes must be produced and the following events must be both possible and coordinated:

- the ovum must be able to travel down the uterine tube
- sperm must be able to swim through the female reproductive tract to reach and fertilise the ovum
- the endometrium must be ready to receive the embryo.

About 10% of couples are infertile, with the most common cause being failure to ovulate. Listed below are some of the causes of infertility that you should be familiar with.

### Failure to ovulate

This is usually the result of a **hormone imbalance** and is treated by using **fertility drugs** that cause the release of FSH and LH from the pituitary gland. Drinking excessive quantities of alcohol, smoking, stress and obesity can be risk factors.

### Low sperm count

Low levels of sperm production are often caused by a **hormone imbalance** and can be treated using **testosterone**, *in vitro* **fertilisation,** or **artificial insemination** using donor sperm. Additional risk factors include smoking, stress and excessive alcohol consumption.

### Blockage of uterine tubes

Uterine tubes can become blocked through **infection** or abnormal tissue growth. Where blockages cannot be surgically removed, *in vitro* **fertilisation** may be an option.

### Failure to implant

For implantation to occur, the monthly changes in the endometrium must be synchronised with changes in the ovary, so that the endometrium is thick enough to receive the embryo when it enters the uterus. **Hormone imbalances** can prevent coordination of the ovarian and uterine changes and can be treated using **fertility drugs**.

### *In vitro* fertilisation

This technique involves giving hormones to the female, causing several ova to be released at ovulation. A syringe is inserted into the female's abdominal cavity, allowing the ova to be collected and placed in liquid nutrient. Sperm are either added to the liquid or injected into an ovum to bring about fertilisation. Several embryos are then inserted into the female's uterus.

### Artificial insemination

In artificial insemination, semen is collected and inserted into the female reproductive tract without intercourse having taken place. When a male has a low sperm count, samples of his sperm can be combined for artificial insemination. Where the male is infertile, sperm from a donor can be used to inseminate the female.

### LET'S THINK ABOUT THIS

Answer the following questions.

1  Explain why mothers are advised to breastfeed their babies, at least for the first few days after birth.

2  Which hormones are found in the contraceptive pill?

3  The most common cause of infertility is failure to ovulate. Describe how this could be treated.

4  Outline possible treatment options for couples in which the male has a low sperm count.

# POSTNATAL DEVELOPMENT

## OVERALL BODY GROWTH

Growth occurs when the body increases in size, and can be measured as changes in height, volume or mass. Growth does not occur at a constant rate throughout life. It is fastest in the first year of life and slows until adolescence, when the onset of puberty starts a second growth spurt, beginning earlier in girls than in boys. Although growth in height usually stops by the end of the teenage years, both males and females continue to increase in weight as muscle development continues.

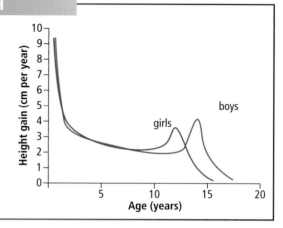

## ORGAN SYSTEM GROWTH

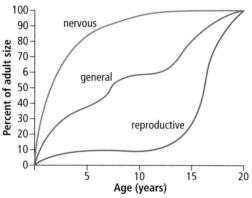

Different parts of the body grow at different rates.

During early development, the brain grows rapidly (important, as it controls the rest of the body), reaching 90% of its adult weight by the age of 5. As a result of this, the head at birth forms about 25% of total height. By adulthood, the head will form only 12% of total height.

The organs of the reproductive system remain relatively small until the teenage years, when at puberty they grow rapidly to reach adult size.

## BODY CHANGES AT PUBERTY

At puberty, under the influence of the sex hormones testosterone (in boys) and oestrogen (in females):

- the rate of growth of the reproductive system increases
- boys and girls develop secondary sex characteristics.

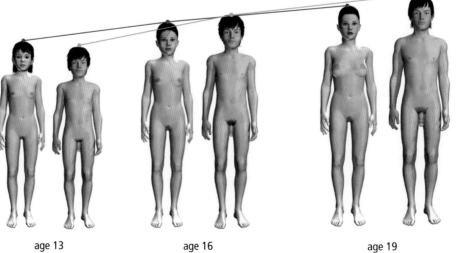

age 13      age 16      age 19

contd

## BODY CHANGES AT PUBERTY contd

The table below shows the changes that take place.

| Body changes at puberty | |
| --- | --- |
| **Males** | **Females** |
| sex organs enlarge<br>sperm production begins in the testes<br>body and facial hair develops<br>skeletal muscles enlarge<br>larynx enlarges causing the voice to deepen<br>sweating increases | sex organs enlarge<br>menstrual cycle begins<br>hips widen and breasts enlarge<br>body hair develops<br>sweating increases |

# HORMONAL CONTROL OF GROWTH

You should be familiar with the role of the following hormones in growth and development.

## Oestrogen, progesterone and testosterone

Release of these sex hormones at puberty causes development of the sex characteristics.

## Growth hormone

Produced by the pituitary gland, growth hormone (GH) acts on bone cells and soft-tissue cells, increasing the rate of amino acid uptake and protein synthesis. As a result, proteins are synthesised rapidly, promoting growth of muscle and bone tissues.

The table below shows the effects of abnormal levels of growth hormone on bone growth.

| Level of production | Effect on bone growth |
| --- | --- |
| under-production of GH during adolescence | overall growth is reduced – individuals have small body size but normal body proportions |
| over-production of GH during adolescence | overall growth is increased – individuals have a large body size but normal body proportions |
| over-production of GH after adolescence | hand, foot and jaw bones are noticeably enlarged |

## Thyroid-stimulating hormone and thyroxine

Thyroid-stimulating hormone (**TSH**) is produced by the pituitary gland and acts on cells in the thyroid gland. The thyroid gland responds by producing other hormones, including **thyroxine**. Thyroxine increases the metabolic rate of body cells, which stimulates growth.

Under-secretion of TSH causes a decrease in thyroxine production and a drop in the metabolic rate. In children, under-secretion can cause a form of dwarfism which is characterised by stunted bone growth and mental disability.

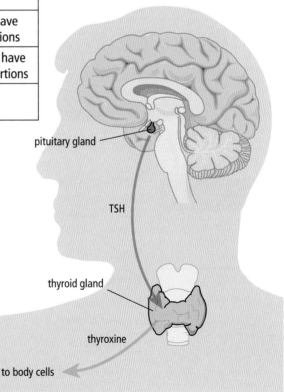

pituitary gland

TSH

thyroid gland

thyroxine

to body cells

## LET'S THINK ABOUT THIS

The level of thyroxine in the blood is controlled by a negative feedback loop. When the level goes too high, thyroxine inhibits the production of TSH by the pituitary gland. This in turn causes a decrease in thyroxine production until normal levels of thyroxine are restored.

# TRANSPORT I

## THE NEED FOR A TRANSPORT SYSTEM

Multi-cellular organisms require a transport system for two main reasons:

- As organisms increase in size, the surface area to volume ratio decreases. The rate of diffusion across the outer surface of the organism becomes insufficient to maintain life.
- As the organism increases in size, the internal distances over which diffusion would have to take place become too great to be efficient.

The human body has two transport systems: the **lymphatic system** and the **blood circulatory system**.

www.wadsworthmedia.com/
biology/0495119814_starr/
big_picture/ch33_bp.html

## LYMPHATIC SYSTEM

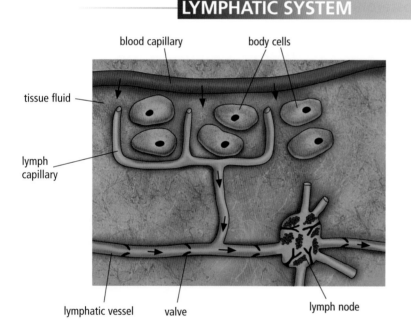

blood capillary    body cells

tissue fluid

lymph capillary

lymphatic vessel    valve    lymph node

Most cells in the human body are not in direct contact with the blood. Instead, they are bathed in tissue fluid, which has passed out through the walls of the blood capillaries, allowing exchange of metabolites between tissue cells and the blood. Some tissue fluid drains back into the bloodstream, but most enters blind-ending tubes called **lymphatic capillaries** to form **lymph**. Lymphatic capillaries drain into larger **lymphatic vessels**, which link up to form the lymphatic system. As there is no pump in this transport system (unlike the blood transport system, which has the heart), flow of lymph is brought about by contraction of muscles in the surrounding tissues. Valves within the lymphatic vessels prevent backflow of fluid.

The largest lymphatic vessels return lymph to the blood circulatory system by draining into two large veins in the chest (the subclavian veins). As lymph flows through the lymphatic vessels, it passes through swellings called **lymph nodes**. These can be found either singly or accumulated in groups in areas such as the groin and the armpits.

Functions of the lymphatic system are as follows.

### Removal of excess tissue fluid

The lymphatic system returns tissue fluid into the blood, maintaining blood volume and preventing excess tissue fluid from building up in the tissues.

### Defence

Lymphocytes and macrophages enter the lymphatic system by squeezing through the walls of the lymphatic vessels. They can be found accumulated in lymph nodes, where they remove invading microbes and other debris before the fluid is returned to the blood.

### Transport of products of fat digestion

Products of fat digestion enter lacteals in the villi of the small intestine. Lacteals are lymphatic capillaries and drain into larger lymphatic vessels in the gut.

> **DON'T FORGET**
>
> Lymph will contain white blood cells but no red blood cells.

contd

LYMPHATIC SYSTEM contd

## Lymph nodes

Within each lymph node, lymphocytes and macrophages lie within a framework of connective tissue fibres. Here, the cells of the immune system filter the lymph, removing invading microbes and any other solid debris that may be present.

http://student.britannica.com/lm/animations/olympha001d4/product.html

# BLOOD CIRCULATORY SYSTEM

The blood circulatory system is a closed system consisting of **blood**, **blood vessels** and the **heart**. Blood is composed of several elements:

- red and white blood cells
- platelets – cell fragments involved with blood clotting
- plasma – mainly water with dissolved nutrients and waste products.

## Red blood cells

Red blood cells are specialised for the **transport of oxygen** around the body.

### Structure

Red blood cells are small and flexible, allowing them to squeeze through capillaries. Their biconcave shape gives a large surface area over which diffusion can take place.

Red blood cells do not contain nuclei. This allows more of the cell's volume to be taken up by **haemoglobin**, a protein that binds with oxygen to form **oxyhaemoglobin**. Lack of a nucleus, however, prevents the cell from making new proteins, limiting its lifespan.

### Manufacture

Red blood cells are made in the red bone marrow of the skull, ribs, pelvis and long bones. **Vitamin $B_{12}$** and **iron** are required for red blood-cell production. To absorb vitamin $B_{12}$ in the gut, **intrinsic factor** (secreted by the stomach) must be present.

### Lifespan

Due to lack of a nucleus and other organelles, a red blood cell can survive only 120 days.

### Breakdown

Old red blood cells are removed by macrophages in the bone marrow, spleen and liver. Haemoglobin is broken down and recycled as follows:

- amino acids are recycled and used to make new proteins
- iron from the haem groups is used to make new red blood cells
- the remainder of the haem group is broken down to form **bilirubin**, which is a constituent of bile.

## LET'S THINK ABOUT THIS

The surface area of a human can be estimated using a nomogram.

To use a nomogram, a line is drawn between an individual's height and mass, and the estimated surface area is read from the middle line. In this example, an individual with a height of 195 cm and mass 58 kg has a surface area of approximately 1·88 m².

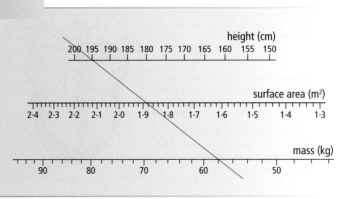

# TRANSPORT II

## THE HEART

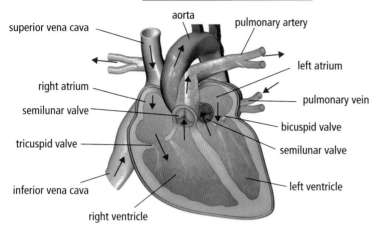

superior vena cava
aorta
pulmonary artery
left atrium
right atrium
pulmonary vein
semilunar valve
bicuspid valve
tricuspid valve
semilunar valve
left ventricle
inferior vena cava
right ventricle

The heart is the muscular organ that pumps blood around the body. It is often referred to as a double pump; the right side of the heart pumps deoxygenated blood to the lungs, and the left side of the heart pumps oxygenated blood to all parts of the body. The heart is made up of four chambers: two **atria** that receive blood from the main veins, and two **ventricles** that pump blood either to the lungs (right ventricle) or to the body (left ventricle). The heart muscle (cardiac muscle) is supplied by the **coronary arteries**. Valves within the heart are present to prevent backflow of blood.

| Name of valve | Location | Phase of cardiac cycle when valve is closed | Function of valve |
|---|---|---|---|
| atrioventricular valves (tricuspid and bicuspid) | between the atria and ventricles | ventricular systole | prevent backflow of blood into the atria |
| semilunar valves | at the start of the pulmonary artery (on the right) and the aorta (on the left) | atrial systole | prevent backflow of blood from the main arteries into the ventricles |

www.sumanasinc.com/webcontent/animations/content/human_heart.html

### DON'T FORGET

To follow the cardiac cycle, you only need to consider one side of the heart (as the right side will be at the same stage as the left side).

http://bcs.whfreeman.com/thelifewire/content/chp49/49020.html

## Cardiac cycle

The sequence of filling and emptying of the heart chambers is called the **cardiac cycle**. During the cardiac cycle, contraction and relaxation of cardiac muscle alters the blood pressure within each of the heart chambers, causing the correct flow of blood through the heart. Blood will always flow from high to low blood pressure unless a valve is closed, preventing blood flow. The cardiac cycle is divided into periods of relaxation (**diastole**) and periods of contraction (**systole**).

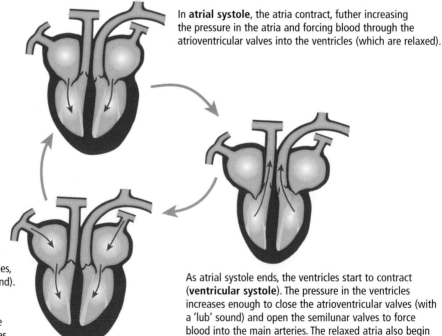

In **atrial systole**, the atria contract, futher increasing the pressure in the atria and forcing blood through the atrioventricular valves into the ventricles (which are relaxed).

In **diastole**, the ventricles relax, causing the pressure to drop below that of the main arteries, closing the semilunar valves (with a 'dup' sound). The atria are relaxed and continue to fill with blood from the vena cava and pulmonary vein, increasing the pressure above that of the ventricles. This forces the atrioventricular valves open, and the ventricles begin to fill.

As atrial systole ends, the ventricles start to contract (**ventricular systole**). The pressure in the ventricles increases enough to close the atrioventricular valves (with a 'lub' sound) and open the semilunar valves to force blood into the main arteries. The relaxed atria also begin to fill from the main veins.

contd

## THE HEART contd

### Control of the cardiac cycle

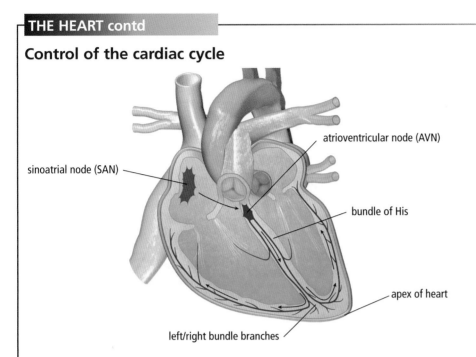

Cardiac muscle is able to beat on its own (it is said to be myogenic). However, to bring about the correct movement of blood, the contraction of each heart chamber must be coordinated. Coordination of the cardiac cycle is brought about by the **conducting system** of the heart.

Electrical excitement (in the form of a **cardiac impulse**) is initiated in an area of the right atrium called the **sinoatrial node (SAN)**, the pacemaker of the heart. From here, a wave of contraction moves out across the atria to reach the **atrioventricular node (AVN)** in the right atrium. The impulse is then passed down through a bundle of fibres in the central wall of the heart to reach the apex of the heart, and then up through left and right branches to the walls of the ventricles. Ventricular contraction (systole) begins at the apex of the ventricles and spreads upwards to squeeze blood out of the ventricles towards the main arteries.

The heart rate is under both nervous and hormonal control:

- **Sympathetic nerves** increase the heart rate.
- **A parasympathetic nerve**, the vagus nerve, slows down the heart rate.
- The part of the brain controlling these nerves is called the **medulla oblongata**.
- **Adrenaline** (a hormone produced by the adrenal glands) works with the sympathetic nerves to speed up the heart rate.

### LET'S THINK ABOUT THIS

What heart rate is shown in the electrocardiogram (ECG) opposite?

Find the time taken for one heart beat by identifying the time between two identical points on the graph, e.g. 0·1 and 0·6 seconds; it takes 0·5 seconds for one heart beat. Every second there will be 2 heart beats, giving a heart rate of:

   2 × 60  =  120 beats per minute

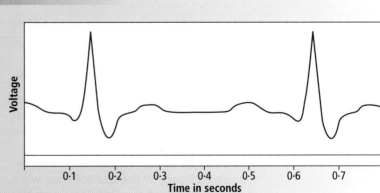

Got to http://www.blaufuss.org/tutonline.html# and click on the SVT tutorial at the bottom.

# TRANSPORT III

## BLOOD VESSELS

The direction of blood flow through the circulatory system is as follows.

### Heart

The heart is the muscular pump that pushes blood around the circulatory system.

### Artery

Arteries carry blood away from the heart. They have thick walls made of elastic fibres and smooth muscle. The elastic fibres allow the vessel to stretch as blood pulses through. This is what you feel when you take your pulse.

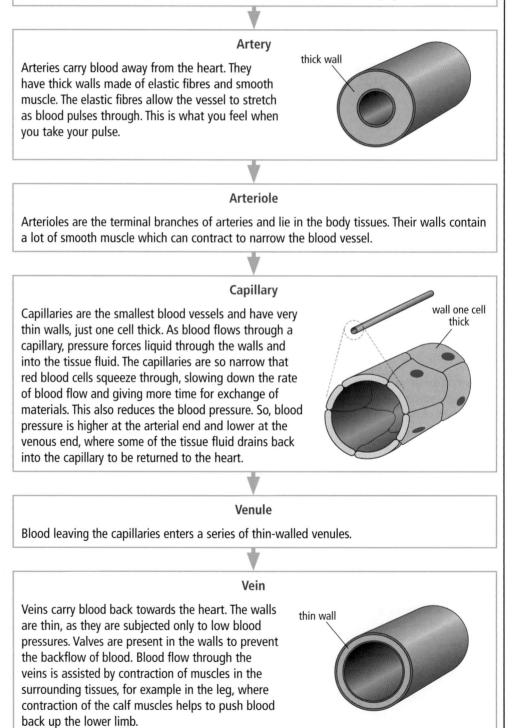

thick wall

### Arteriole

Arterioles are the terminal branches of arteries and lie in the body tissues. Their walls contain a lot of smooth muscle which can contract to narrow the blood vessel.

### Capillary

Capillaries are the smallest blood vessels and have very thin walls, just one cell thick. As blood flows through a capillary, pressure forces liquid through the walls and into the tissue fluid. The capillaries are so narrow that red blood cells squeeze through, slowing down the rate of blood flow and giving more time for exchange of materials. This also reduces the blood pressure. So, blood pressure is higher at the arterial end and lower at the venous end, where some of the tissue fluid drains back into the capillary to be returned to the heart.

wall one cell thick

### Venule

Blood leaving the capillaries enters a series of thin-walled venules.

### Vein

Veins carry blood back towards the heart. The walls are thin, as they are subjected only to low blood pressures. Valves are present in the walls to prevent the backflow of blood. Blood flow through the veins is assisted by contraction of muscles in the surrounding tissues, for example in the leg, where contraction of the calf muscles helps to push blood back up the lower limb.

thin wall

## NAMED VESSELS OF THE CIRCULATORY SYSTEM

You should be able to name the vessels in the following diagram.

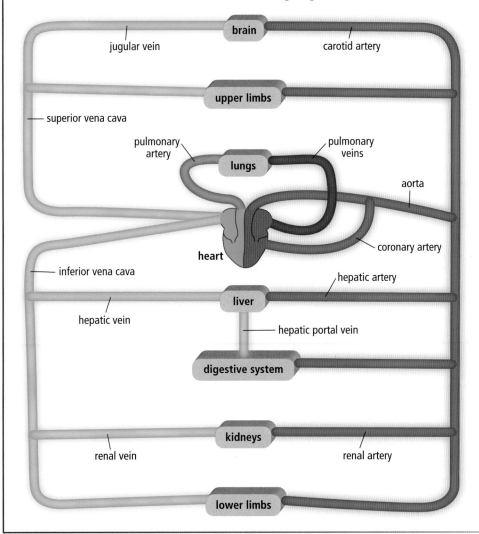

**DON'T FORGET**

The pulmonary artery carries deoxygenated blood. The pulmonary vein carries oxygenated blood.

## LET'S THINK ABOUT THIS

Answer the following questions.

1  Name the blood vessel carrying oxygenated blood from the lungs to the heart.

2  Describe the blood supply to the liver.

3  Explain the function of (**i**) elastic fibres in artery walls (**ii**) smooth muscle fibres in arteriole walls.

4  Which blood vessels have valves in their walls? What is their function?

5  Make a list of the changes in composition of the blood that occur as blood flows from the arterial end to the venous end of a capillary.

# TRANSPORT IV

**DON'T FORGET**

You should remember that blood always flows from high to low pressure, unless a closed valve prevents blood flow.

## BLOOD PRESSURE

### Blood pressure in the heart

The diagram below shows the changes in pressure that occur during the cardiac cycle in the left atrium, left ventricle and the aorta.

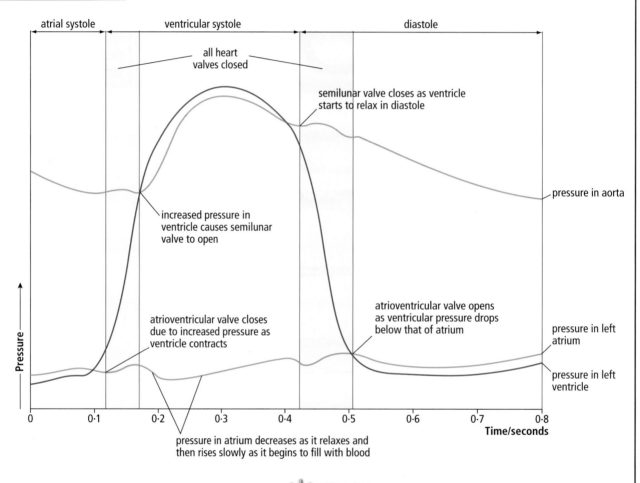

atrial systole | ventricular systole | diastole

all heart valves closed

semilunar valve closes as ventricle starts to relax in diastole

increased pressure in ventricle causes semilunar valve to open

atrioventricular valve closes due to increased pressure as ventricle contracts

atrioventricular valve opens as ventricular pressure drops below that of atrium

pressure in aorta

pressure in left atrium

pressure in left ventricle

Pressure →

Time/seconds

pressure in atrium decreases as it relaxes and then rises slowly as it begins to fill with blood

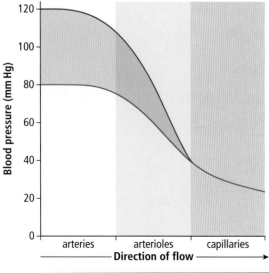

Blood pressure (mm Hg)

arteries | arterioles | capillaries

**← Direction of flow →**

http://bcs.whfreeman.com/thelifewire/content/chp49/49020.html

### Blood pressure in the blood vessels

In the blood vessels, blood pressure is caused by the pumping action (contraction and relaxation) of the heart ventricles. During ventricular systole, blood pressure is at its highest (about 120 mm Hg), decreasing to about 80 mm Hg during ventricular diastole.

In the arteries, the walls bulge during systole as a wave of blood passes through, recoiling during diastole and pushing blood through. As blood enters narrower vessels, the resistance of the vessel walls increases, causing a decrease in blood pressure. This decrease is greatest in the arterioles, which have the largest resistance. Blood pressure continues to decrease as blood flows through capillaries, venules and veins.

# OXYGEN DISSOCIATION CURVES

Oxygen dissociation curves demonstrate the uptake and release of oxygen by haemoglobin. The graph plots the percentage saturation (the percentage of haemoglobin that is bound to oxygen, forming oxyhaemoglobin) against partial pressure (the pressure exerted by oxygen). The partial pressure of oxygen is directly proportional to the amount of oxygen present; as the partial pressure increases, the amount of oxygen present increases.

In the lungs, where the partial pressure of oxygen is high, haemoglobin binds to oxygen to form oxyhaemoglobin. As the blood passes through the body tissues, the partial pressure of oxygen decreases, and oxyhaemoglobin dissociates to release oxygen. The oxygen is then available for tissue respiration.

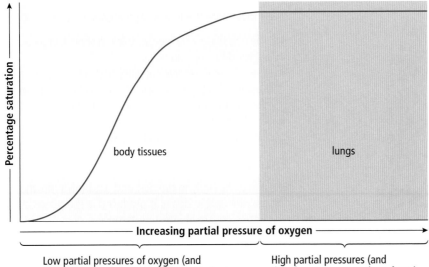

Low partial pressures of oxygen (and therefore less oxygen) are found in the body tissues. Here, haemoglobin is less able to bind to oxygen and releases it to body cells.

High partial pressures (and therefore more oxygen) are found in the lungs, where almost 100% of haemoglobin binds to oxygen.

## Effect of temperature

As temperature increases, the affinity of haemoglobin for oxygen decreases. Therefore, when body tissues are warmer, more oxyhaemoglobin dissociates and more oxygen is released from the blood to the body cells.

During **exercise**, the rate of respiration in skeletal muscle cells increases and the cells generate more heat energy, causing an increase in muscle temperature. The increased temperature results in more efficient unloading of oxygen from oxyhaemoglobin, and more oxygen becomes available for respiration.

After tissue damage through either **injury** or **infection**, the localised area becomes warmer. The blood releases more oxygen to the tissue, allowing increased respiration which facilitates a faster repair.

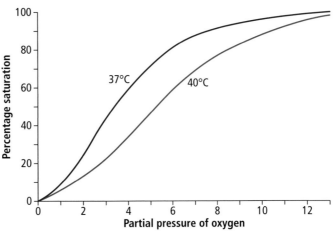

# LET'S THINK ABOUT THIS

Foetal haemoglobin has a higher affinity for oxygen than adult haemoglobin. This allows foetal blood to pull oxygen across the placenta. Where would the oxygen dissociation curve for foetal haemoglobin fit on the graph above?

# FATE OF ABSORBED SUBSTANCES AND THE LIVER

## FATE OF ABSORBED FOOD MOLECULES

Fats, carbohydrates and proteins are broken down in the gut and are absorbed through the wall of the small intestine. The small intestine has a large surface area to maximise the rate of absorption of these molecules. In particular, the inner wall has finger-like processes called **villi**.

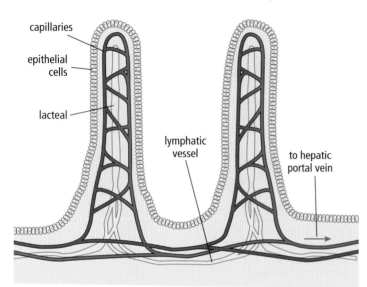

capillaries

epithelial cells

lacteal

lymphatic vessel

to hepatic portal vein

You should note the following features of villus structure:

- The lining layer is one cell thick, a short distance over which diffusion can occur.
- There is an extensive capillary bed, allowing rapid absorption of glucose and amino acids. Blood leaving the villi eventually enters the hepatic portal vein.
- The **lacteal** is a lymphatic capillary. Products of fat digestion enter the lymphatic system here.

### Fats

Fats enter the body in the diet and are broken down by the enzyme lipase. Lipase is produced in the pancreas and enters the small intestine through a duct. Here, fats are broken down into fatty acids and glycerol. In the small intestine, food is mixed with bile. Produced by the liver and stored in the gall bladder, bile is released into the small intestine after we eat. Bile emulsifies fat, increasing the surface area available for lipase action.

Products of fat digestion are absorbed by lacteals in the villi and travel through the lymphatic system before entering the blood. Fat can be used as a respiratory substrate, with excess fat being stored under the skin and around body organs.

**DON'T FORGET**

Bile does not break down fat molecules – the molecules simply spread out more.

### Carbohydrates

Carbohydrates enter the body in the diet and are broken down by several enzymes:

- Starch is broken down into maltose by salivary and pancreatic amylase.
- Maltose is broken down into glucose by maltase produced in the small intestine.

Glucose enters the blood capillaries in the villi and passes through the **hepatic portal vein** to the liver. It is the main respiratory substrate and can be converted to the storage carbohydrate **glycogen** in both the liver and the muscles.

### Proteins

Proteins in the diet are broken down by a group of enzymes called proteases. You should be familiar with the proteases pepsin and trypsin. Proteins are broken down into amino acids which are absorbed by the blood capillaries in the villi and pass through the hepatic portal vein to the liver. Amino acids circulate in the blood and are used to build new proteins in cells.

Excess amino acids are broken down in the liver (a process called **deamination**) to produce **urea**.

## THE LIVER

www.britannica. com/EBchecked/ topic/344579/liver

The liver acts as a chemical factory; it builds up large molecules from substances absorbed by the small intestine and breaks down toxic molecules in the bloodstream. It lies in the upper abdomen, below the diaphragm. Just below the liver lies the **gall bladder**, which acts as a temporary store for bile.

contd

## THE LIVER contd

### Blood supply

The liver has a dual blood supply:

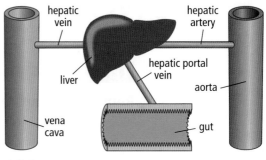

- The **hepatic artery** (a branch of the aorta) brings oxygenated blood to the liver.
- The **hepatic portal vein** carries deoxygenated blood rich in dissolved foods (amino acids, glucose and vitamins) from the stomach and small intestine.

Venous blood leaves the liver through the **hepatic vein** and enters the inferior vena cava.

The relative concentrations of substances in the blood vessels associated with the liver are shown below.

| | Blood vessel | | |
|---|---|---|---|
| | **Hepatic artery** | **Hepatic portal vein** | **Hepatic vein** |
| **Oxygen concentration** | high | normal | normal |
| **Carbon dioxide concentration** | normal | high | high |
| **Glucose concentration** | normal | high | normal |
| **Amino acid concentration** | normal | high | normal |
| **Urea concentration** | normal | normal | high |

### Liver functions

The liver has a wide range of functions including bile production, conservation of useful substances (see table below), detoxification (see table at bottom of page) and heat production.

| Substance | Role of the liver in conservation of substance |
|---|---|
| carbohydrate | When the blood glucose level increases after eating a meal, cells in the pancreas secrete the hormone **insulin**. Insulin activates liver enzymes which convert glucose into glycogen, bringing the blood glucose level back to normal. When the blood sugar level drops below normal, release of **glucagon** from the pancreas activates liver enzymes to convert glycogen to glucose, raising the blood glucose level. |
| fat | Within the liver, lipid molecules are altered to form:<br>• phospholipids and cholesterol (both used in cell membranes)<br>• lipoproteins (for lipid transport)<br>• bile salts (to emulsify lipids)<br>• steroid hormones (such as sex hormones). |
| protein | Liver cells manufacture plasma proteins, such as **fibrinogen** which is important in clotting of blood. |
| vitamins and minerals | The liver acts as a temporary store for some vitamins (for example, A, $B_{12}$, D, E) and minerals (such as potassium, copper and iron). |

The liver can act as a filter which removes or alters toxic substances in the blood, by either inactivation or degradation.

| Method of removal | Substance to be detoxified |
|---|---|
| inactivation | Some hormones, such as sex hormones and insulin, are inactivated in the liver before being excreted in bile or urine. |
| degradation | Liver cells can break down some substances to make them less toxic. This includes substances that are products of metabolism, as well as substances that have been ingested, such as drugs and alcohol. |

### LET'S THINK ABOUT THIS

What changes would you expect in the concentrations of urea and glucose as blood flows through the liver from the hepatic portal vein to the hepatic vein (**i**) after a heavy meal and (**ii**) during fasting?

# THE LUNGS AND THE KIDNEYS

## THE LUNGS

### Removal of carbon dioxide

Carbon dioxide is a waste product of respiration. It diffuses out of body cells and enters the blood, where it is transported in the plasma. In the lungs, it diffuses out of the capillaries and into the air in the alveoli. Air rich in carbon dioxide is then exhaled.

## THE KIDNEYS

The kidneys are paired organs located on the posterior abdominal wall. They receive blood containing excess water, urea and other waste substances through the renal artery. Blood is filtered in the kidney and leaves through the renal vein. The functional unit of the kidney is the **nephron**.

The function of each part of the nephron is shown below.

### Glomerulus
The branches of the renal artery eventually supply a knot of capillaries (a **glomerulus**). The glomerulus is the site of **ultrafiltration**. The walls of the glomerulus contain pores that let small molecules such as sodium, water, glucose and urea pass through into the **Bowman's capsule**. The blood vessel entering the glomerulus is wider than the vessel which leaves it, producing the high pressure that is required for ultrafiltration.

### Bowman's capsule
The Bowman's capsule is a cup-shaped hollow structure which receives the filtrate after it leaves the blood.

### Proximal convoluted tubule
Useful molecules are reabsorbed from the proximal convoluted tubule. The cells lining this part of the nephron are adapted for active transport in the following ways:

- the surface area is increased by the presence of microvilli
- many mitochondria are present to provide ATP
- carrier proteins are present to transport glucose across the membrane.

Amino acids, glucose, sodium and chloride ions are actively absorbed here. Due to this active transport, the water concentration of the blood plasma is reduced, and water passes into the blood by osmosis. About 85% of water is reabsorbed from the proximal convoluted tubule.

### Loop of Henlé
Salts and water are reabsorbed here. Reabsorption in the ascending part of the loop of Henlé is influenced by the hormone ADH.

### Distal convoluted tubule
In the distal convoluted tubule, water is reabsorbed under the influence of ADH.

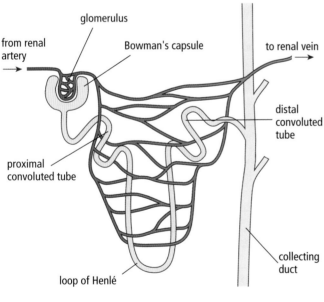

### Collecting duct
Filtrate from the nephrons passes into the collecting ducts. Some water is reabsorbed here, under the influence of ADH. The collecting ducts transport the newly formed urine to the ureter.

contd

## THE KIDNEYS contd

# Hormonal control of water concentration

The process of maintaining water and salt balance in the body is called **osmoregulation**. Osmoregulation uses negative feedback to maintain a constant water concentration in the body. The receptor cells (**osmoreceptors**) that detect the water concentration of the blood are found lining the blood vessels of the **hypothalamus**, the monitoring centre in the brain. From the osmoreceptors, nerve impulses pass to the pituitary gland, stimulating or inhibiting the production of **antidiuretic hormone (ADH)** from the pituitary gland.

In the case of a low water concentration in the blood, ADH production is stimulated and the hormone travels in the blood to the kidney tubules. Here, it acts on the ascending limb of the loop of Henlé, the distal convoluted tubule and the collecting ducts, making them more permeable to water. Water passes out of these structures by osmosis and into the blood in the surrounding capillaries. If more ADH is produced, less urine is excreted.

If there is a high water concentration in the blood, less ADH is released by the pituitary gland, resulting in more water being excreted.

The diagram below summarises the events of osmoregulation.

www.bbc.co.uk/
schools/gcsebitesize/
science/add_aqa/
homeo
www.sumanasinc.
com/webcontent/
animations/content/
kidney.html

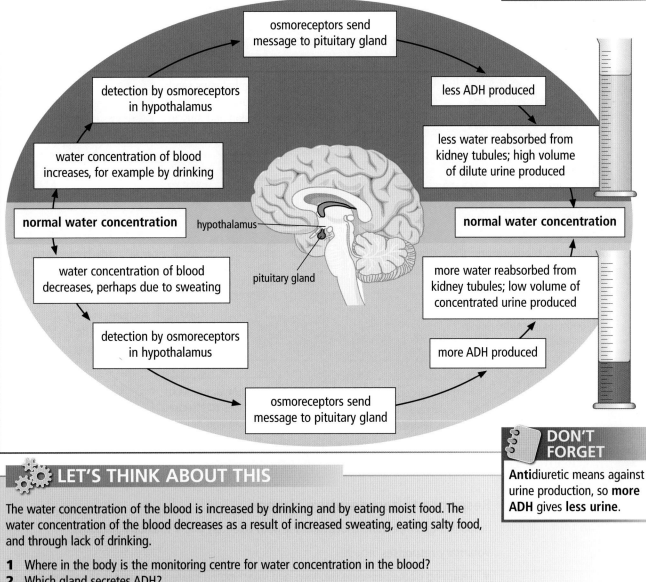

osmoreceptors send message to pituitary gland

detection by osmoreceptors in hypothalamus

less ADH produced

water concentration of blood increases, for example by drinking

less water reabsorbed from kidney tubules; high volume of dilute urine produced

**normal water concentration**

hypothalamus

**normal water concentration**

water concentration of blood decreases, perhaps due to sweating

pituitary gland

more water reabsorbed from kidney tubules; low volume of concentrated urine produced

detection by osmoreceptors in hypothalamus

more ADH produced

osmoreceptors send message to pituitary gland

## LET'S THINK ABOUT THIS

### DON'T FORGET

**Anti**diuretic means against urine production, so **more ADH** gives **less urine**.

The water concentration of the blood is increased by drinking and by eating moist food. The water concentration of the blood decreases as a result of increased sweating, eating salty food, and through lack of drinking.

1 Where in the body is the monitoring centre for water concentration in the blood?
2 Which gland secretes ADH?
3 How does ADH reach the kidney tubules?
4 What is the effect of increased ADH secretion?
5 Explain why the concentration of urea increases as the filtrate flows from the Bowman's capsule to the collecting ducts.

# HOMEOSTASIS I

Homeostasis is the maintenance of the body's internal environment in response to changes in the surroundings. This includes:

- water balance (see page 63)
- temperature
- blood sugar level.

## TEMPERATURE REGULATION

All chemical reactions in the body are controlled by enzymes. When body temperature is below the optimum temperature for enzyme function, metabolism is slow. Above the optimum, enzymes start to denature; the metabolism slows down and eventually stops.

Receptors that detect blood temperature (thermoreceptors) are found in the lining of blood vessels in the **hypothalamus** – the temperature monitoring centre in the brain. When thermoreceptors detect changes in blood temperature, the hypothalamus sends out nerve impulses to effector organs in the skin and body muscles. The effectors bring about a response designed to return the temperature to normal. The diagram below summarises the process.

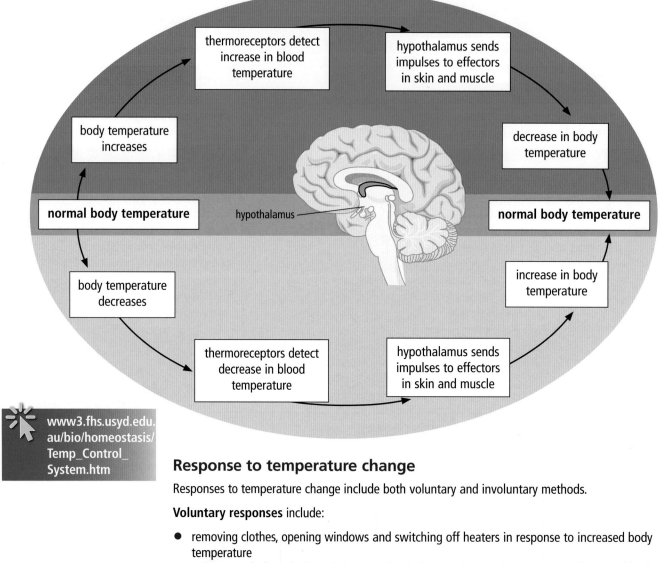

www3.fhs.usyd.edu.au/bio/homeostasis/Temp_Control_System.htm

## Response to temperature change

Responses to temperature change include both voluntary and involuntary methods.

**Voluntary responses** include:

- removing clothes, opening windows and switching off heaters in response to increased body temperature
- putting on clothes, closing windows and switching on heaters in response to decreased body temperature.

**Involuntary responses** to temperature change are shown opposite.

contd

## TEMPERATURE REGULATION contd

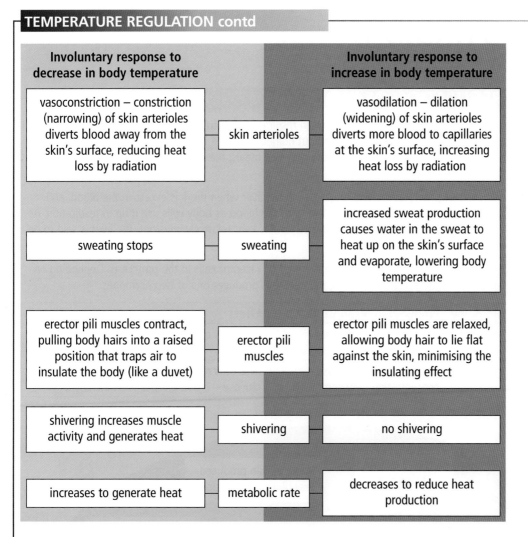

| Involuntary response to decrease in body temperature | | Involuntary response to increase in body temperature |
| --- | --- | --- |
| vasoconstriction – constriction (narrowing) of skin arterioles diverts blood away from the skin's surface, reducing heat loss by radiation | skin arterioles | vasodilation – dilation (widening) of skin arterioles diverts more blood to capillaries at the skin's surface, increasing heat loss by radiation |
| sweating stops | sweating | increased sweat production causes water in the sweat to heat up on the skin's surface and evaporate, lowering body temperature |
| erector pili muscles contract, pulling body hairs into a raised position that traps air to insulate the body (like a duvet) | erector pili muscles | erector pili muscles are relaxed, allowing body hair to lie flat against the skin, minimising the insulating effect |
| shivering increases muscle activity and generates heat | shivering | no shivering |
| increases to generate heat | metabolic rate | decreases to reduce heat production |

## Hypothermia

When the body's temperature falls to subnormal levels (below 35°C), hypothermia results. Adults who are exposed to extreme conditions, particularly if clothing becomes wet, may experience hypothermia. In normal conditions, hypothermia is more common in infants and the elderly, who are less able to produce and retain heat.

### Infants

Heat is lost through the surface of the body to the surrounding air. As babies have a larger surface-area-to-volume ratio than adults, they can lose heat rapidly. In addition, regulatory mechanisms are not fully developed in babies:

- they are incapable of using voluntary methods (such as putting on clothes)
- the nervous system is not mature enough to provide involuntary control.

### Elderly people

The elderly are at risk of hypothermia for several reasons:

- levels of physical activity are decreased, so less heat is generated by muscular activity
- their metabolic rates are decreased, resulting in less heat production
- temperature regulatory mechanisms function less efficiently
- they may not be able to afford to heat their homes.

## LET'S THINK ABOUT THIS

Changes in surface temperature on the body can be measured using a thermistor. Use your class notes to review this experiment.

# HOMEOSTASIS II

## REGULATION OF BLOOD GLUCOSE LEVEL

Glucose is used by body cells as the main respiratory substrate, and a constant supply is required to give energy in the form of ATP. Without homeostasis, the blood glucose level would be very high after a meal and become very low between meals; the required steady supply of energy could not be achieved.

Homeostasis promotes storage of glucose in the liver when there is excess in the blood, and stimulates release of glucose by the liver into the blood as body cells use it up in respiration. As a result of homeostasis, the blood glucose level is kept relatively constant. The liver is said to be a storage reservoir for carbohydrates.

The blood glucose concentration is monitored by receptor cells in the **pancreas**. Depending on the glucose concentration detected, the pancreas produces one of two hormones.

| Hormone | Produced in response to | Effect |
|---------|------------------------|--------|
| insulin | increased blood glucose concentration after eating a meal | liver cells respond by converting glucose into glycogen – blood glucose level decreases |
| glucagon | decreased blood glucose concentration between meals | liver cells respond by converting glycogen into glucose – blood glucose level increases |

The diagram below summarises the steps involved.

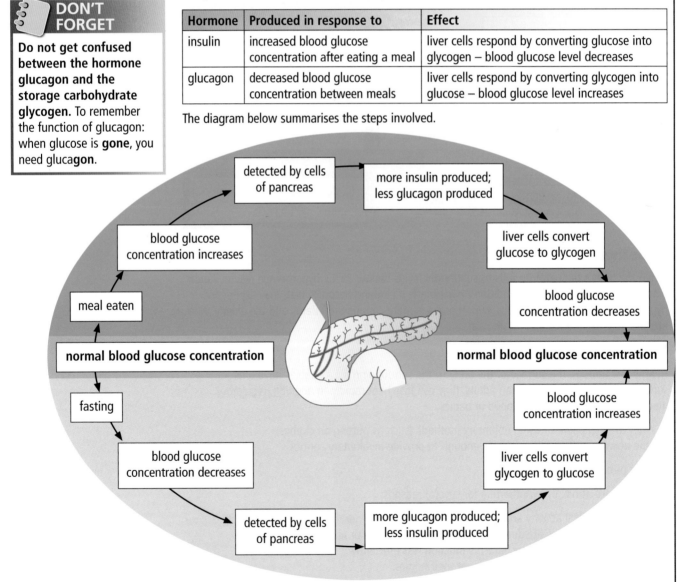

## Adrenaline

The hormone **adrenaline** is produced by the adrenal glands in times of stress and is involved in preparing the body for the 'flight or fight response'. In the presence of adrenaline, liver cells convert glycogen into glucose, increasing the blood glucose concentration above the normal level. This provides muscles with the glucose they require during increased levels of activity.

contd

## REGULATION OF BLOOD GLUCOSE LEVEL contd

### Diabetes

In diabetes, the pancreatic cells that produce **insulin** do not function properly. As a result, the blood glucose concentration rises so much that reabsorption in the kidneys is incomplete and some glucose is excreted in the urine.

Diabetes is diagnosed using a **glucose tolerance test**. Here, the individual being tested fasts for 8 hours before consuming a known mass of glucose. The blood glucose level is then measured at regular intervals.

In an individual without diabetes, the glucose level rises after consumption and then falls rapidly as insulin is produced.

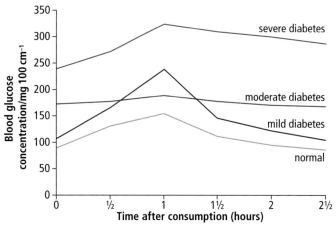

People with **mild diabetes** have relatively normal fasting blood glucose levels. However, after consumption, insulin production is not great enough to remove glucose quickly, and glucose levels become much higher than in a normal individual. As some insulin is produced, blood glucose falls to the fasting level during the test period but at a low rate.

In **severe diabetes**, there is little or no insulin production. As a result, fasting blood glucose levels are much higher than normal. After consumption, they increase further and do not return to fasting level during the test.

## EFFECTS OF EXERCISE ON HEART RATE AND LUNGS

When muscle cells are active during exercise, their respiration rate increases to provide the additional ATP required to bring about contraction. To provide additional oxygen and glucose, and to allow the removal of the increased levels of carbon dioxide and heat that are produced, sympathetic nerves bring about the following changes:

- the **cardiac output** increases due to increases in both the volume of blood being pumped by the heart (stroke volume) and the heart rate
- blood is redistributed in the body, increasing the supply to the muscles and skin, and decreasing the supply to organ systems not involved in exercise, for example the digestive system
- the breathing rate and volume of air inhaled with each breath increase.

## LET'S THINK ABOUT THIS

The diagram below shows some of the stages in the control of a person's blood glucose level.

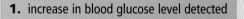

**1.** increase in blood glucose level detected

**2.** hormone X released

**3.** glucose converted to substance Y

**4.** blood glucose level decreases

**(a)** In which organ would the increase in blood glucose level be detected?

**(b)** Name hormone X and substance Y.

**(c)** Where does stage 3 occur?

**(d)** If hormone X was not produced, the blood glucose level would drop very slowly. Why?

# NERVOUS SYSTEM I

## NERVE CELLS

The functional cell of the nervous system is the nerve cell or **neurone**. There are three types of neurone: **sensory**, **motor** and **relay**. Each neurone consists of three parts.

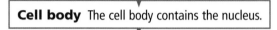

**Dendrite** Dendrites receive nerve impulses and send them towards the cell body.

**Cell body** The cell body contains the nucleus.

**Axon** The axon carries the nerve impulse away from the cell body.

### Sensory neurones

Sensory neurones pass information from **sense receptors** to neurones in the central nervous system (CNS). A single dendrite receives information from sense receptors and transmits a nerve impulse towards the cell body. From the cell body, a single axon carries the nerve impulse into the spinal cord where the impulse is transmitted to the dendrites of a relay neurone.

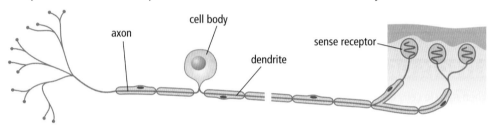

### Relay neurones

Relay neurones lie completely within the CNS. They vary in shape but, in general, have several dendrites and an axon extending from the cell body, allowing messages to be passed on to a large number of other neurones.

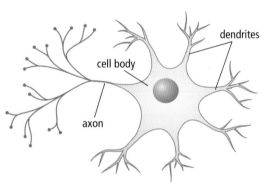

### Motor neurones

Motor neurones transmit nerve impulses from the CNS to an effector organ (muscle or gland). The cell body and several short dendrites lie embedded in the CNS. One axon extends from the cell body, passing out of the CNS to reach the effector organ.

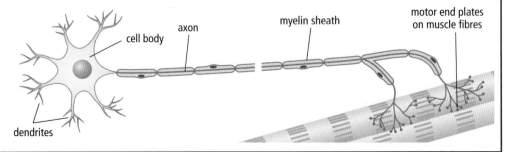

## THE SYNAPSE

A synapse is the junction between an axon of one cell and a dendrite of another. The synapse allows the transmission of nerve impulses (and therefore communication) between neurones.

The nerve impulse passes down the axon to reach the **axon bulb**, where it stimulates **synaptic vesicles** to move towards and fuse with the **pre-synaptic membrane**, releasing a neurotransmitter chemical by exocytosis. The neurotransmitter molecules diffuse across the **synaptic gap** to reach the **post-synaptic membrane**, where they fuse with receptors in the membrane. If a large enough number of receptors fuse with neurotransmitter, the post-synaptic membrane will reach its **threshold**, and a nerve impulse will be transmitted by the post-synaptic cell.

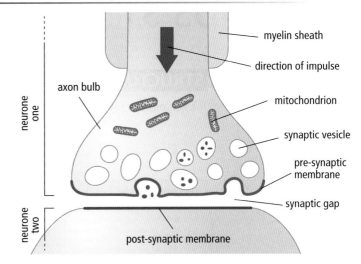

So that the next nerve impulse can be transmitted, neurotransmitters must be removed rapidly from the post-synaptic membrane. If this did not occur, it would be impossible to control the frequency of nerve impulses, and an individual would not be able to distinguish between stronger and weaker stimuli (such as bright and dim light).

http://www.youtube.com/watch?v=HXx9qlJetSU

### Neurotransmitter substances

You should be familiar with the following neurotransmitter substances.

#### Acetylcholine

Acetylcholine is removed from the post-synaptic membrane by enzymatic degradation. The enzyme acetylcholinesterase breaks down the neurotransmitter, and the inactive products are reabsorbed by the pre-synaptic cell. They are then recycled and used to make more acetylcholine.

#### Noradrenaline

Noradrenaline leaves the post-synaptic membrane and is reabsorbed intact by the pre-synaptic cell. Within the pre-synaptic cell, noradrenaline is enclosed within vesicles for reuse.

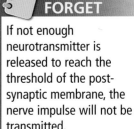

**DON'T FORGET**

If not enough neurotransmitter is released to reach the threshold of the post-synaptic membrane, the nerve impulse will not be transmitted.

## MYELINATION

The speed of conduction of a nerve impulse through a neurone is increased by the presence of a myelin sheath. Myelin is formed by supporting cells which wrap themselves round and round axons, building up layer upon layer of cell membrane.

Myelination is not complete at birth but continues as the child grows, being completed in the upper limbs before the lower limbs. As a result, babies cannot coordinate their movements at birth, and gain control of their arms before their legs.

**DON'T FORGET**

The myelin sheath works just like the plastic insulation around an electrical cable.

## LET'S THINK ABOUT THIS

Explain how (**i**) release of noradrenaline is triggered and (**ii**) how it is removed after transmission of the nerve impulse.

# NERVOUS SYSTEM II

## STRUCTURAL DIVISIONS OF THE NERVOUS SYSTEM

The nervous system consists of the **brain**, **spinal cord** and **peripheral nerves**.

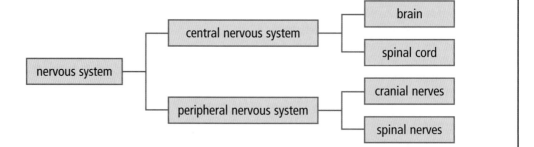

## THE CENTRAL NERVOUS SYSTEM

The central nervous system consists of the brain and spinal cord, and contains neurones (cell bodies, axons and dendrites) and their supporting cells. The tissue that makes up the CNS is divided into white and grey matter.

### White matter

Composed mainly of axons and dendrites, white matter is white in appearance due to myelination of the nerve fibres. It is located centrally within the brain and on the periphery of the spinal cord.

### Grey matter

Grey matter is made up mainly of nerve cell bodies. In the brain, it is located mainly on the outer edges, and in the spinal cord it lies centrally.

## THE BRAIN

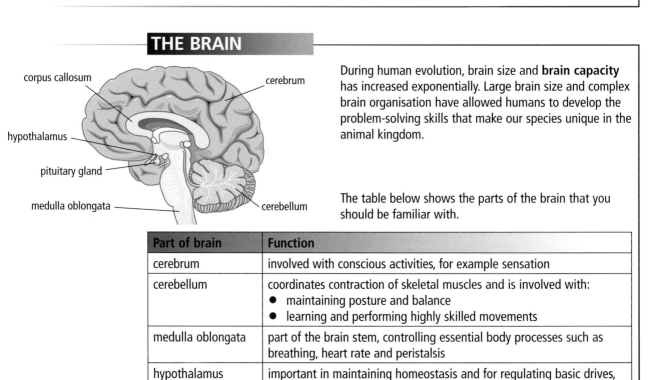

During human evolution, brain size and **brain capacity** has increased exponentially. Large brain size and complex brain organisation have allowed humans to develop the problem-solving skills that make our species unique in the animal kingdom.

The table below shows the parts of the brain that you should be familiar with.

| Part of brain | Function |
|---|---|
| cerebrum | involved with conscious activities, for example sensation |
| cerebellum | coordinates contraction of skeletal muscles and is involved with:<br>● maintaining posture and balance<br>● learning and performing highly skilled movements |
| medulla oblongata | part of the brain stem, controlling essential body processes such as breathing, heart rate and peristalsis |
| hypothalamus | important in maintaining homeostasis and for regulating basic drives, such as sexual behaviour, drinking and eating |

contd

## THE BRAIN contd

### Cerebrum

The cerebrum consists of two cerebral hemispheres (left and right) which are connected by a bridge of nerve fibres called the **corpus callosum**. The corpus callosum is the only route of communication between the hemispheres.

The surface of the cerebrum is folded, allowing more cell bodies to be present. As a result, the number of interconnections between neurones is increased.

### Functional areas of the cerebrum

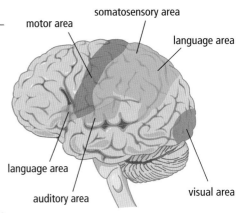

somatosensory area
motor area
language area
language area
auditory area
visual area

**Somatosensory area**
This receives sensory information from the internal organs, skin and muscles. Parts of the body can be mapped out along the somatosensory area, with body parts that experience fine sensation (such as the lips and fingertips) covering a larger area of the cerebrum.

**Motor area**
The motor area controls the contraction of specific muscles. Just like the somatosensory area, the parts of the body are organised along the motor area. Where muscles require fine control (such as muscles of the lips, tongue and fingers), they take up a greater proportion of the motor area. A motor homunculus (shown opposite) can be drawn to show the relative size of the motor area that controls each part of the body.

**Visual area**
The visual area receives information from the eyes through the optic nerves. It interprets colour, shape and movement.

**Auditory area**
The auditory area receives information from the cochlea in the inner ear, through the auditory nerve. It interprets both pitch and rhythm.

**Language areas**
The language areas of the cerebrum are involved with controlling muscles required for speech (for example, muscles of the lips, tongue and larynx) and memory of vocabulary.

**Association areas**
Association areas are concerned with emotions, personality, intelligence and creativity.

www.pbs.org/wgbh/aso/tryit/brain/probe.html

### LET'S THINK ABOUT THIS

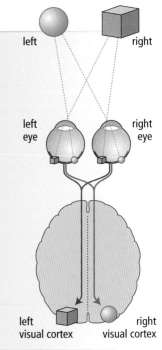

left
right

In split-brain patients, the **corpus callosum** has been cut, often to reduce epileptic seizures, preventing transfer of information between the right and left sides of the cerebrum. To understand the effect that this has on the patient, we must remember:

- Visual information in the left field of view is projected onto the right visual cortex, and information in the right field of view is projected to the left visual cortex.
- The left motor cortex controls muscles on the right side of the body, and the right motor cortex controls muscles on the left side of the body.
- The speech area is usually found only on the left cerebral hemisphere.

left eye
right eye

In the experiment shown opposite, a picture of a ball is in the left field of view, and a cube is in the right field of view. If asked what he has seen, the patient will say 'cube', as this is projected to the left hemisphere – communication between left visual area and speech area takes place. He cannot say 'ball' because communication between the right and left hemispheres has been cut.

If asked to use his left hand to pick up the object he was shown from several hidden under a cloth, he would pick up the ball (communication between right visual and motor areas takes place) but he wouldn't be able to tell you what he was holding.

left
visual cortex
right
visual cortex

http://www.youtube.com/watch?v=ZMLzP1VCANo

# NERVOUS SYSTEM III

## FUNCTIONAL DIVISIONS OF THE PERIPHERAL NERVOUS SYSTEM

The peripheral nervous system can be divided into two functional parts.

### Somatic nervous system

The **somatic system** is responsible for:

- voluntary control of skeletal muscles
- involuntary reflexes.

### Autonomic nervous system

The **autonomic nervous system** regulates the internal environment (for example, heart rate, body temperature, digestion). It has two divisions: **sympathetic** and **parasympathetic**. These perform opposing functions (and are therefore **antagonistic**), the sympathetic division preparing the body for action and the parasympathetic division returning the body to the resting state. For example, sympathetic nerves speed up the heart rate, while a parasympathetic nerve (the vagus nerve) slows down the heart rate (see page 55). Similarly, sympathetic nerves slow down the rate of peristalsis, and parasympathetic nerves speed it up again once the period of excitement is over.

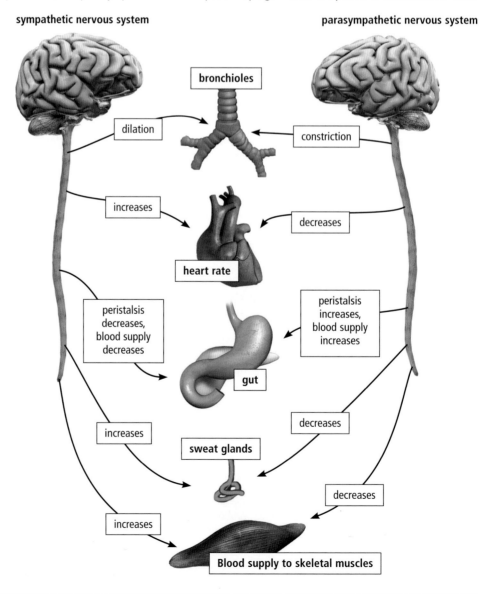

sympathetic nervous system     parasympathetic nervous system

bronchioles — dilation / constriction

increases / decreases — heart rate

peristalsis decreases, blood supply decreases / peristalsis increases, blood supply increases — gut

increases / decreases — sweat glands

increases / decreases — **Blood supply to skeletal muscles**

# CONVERGING AND DIVERGING NEURAL PATHWAYS

The route that a nerve impulse follows as it travels through the nervous system is called a **neural pathway**. Neural pathways can be very complex, with information from several sources converging on the same neurone, or nerve impulses from one area of the brain being sent to different destinations. You should be familiar with two types of neural pathway: converging and diverging neural pathways.

## Converging neural pathways

In a **converging neural pathway**, nerve impulses originating from different sources are directed to one neurone. This is of functional significance, as it allows weak stimuli to be amplified, as shown in the visual pathway. Within the retina, the rods that detect low light levels each release only a small amount of neurotransmitter into the synaptic gap. However, a cumulative effect results, as several rods have synapses with the same post-synaptic neurone, resulting in the transmission of a nerve impulse at low light levels.

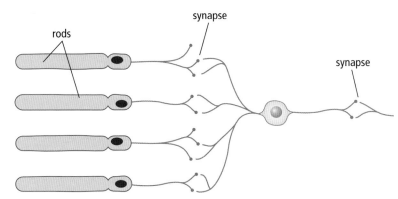

## Diverging neural pathways

In a diverging pathway, a single pre-synaptic neurone forms synapses with several post-synaptic neurones, allowing the pathway to branch, sending information to several destinations at the one time. Diverging neural pathways allow fine motor control, where several skeletal muscles work together to produce a precise movement, for example of the fingers or the eyes.

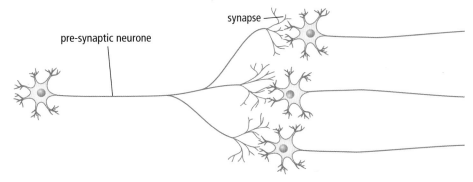

Temperature regulation by the hypothalamus also involves a diverging neural pathway, controlling sweat glands, erector pili muscles, blood vessels and shivering (see page 65).

# PLASTICITY

Functional flexibility (plasticity) of the nervous system allows us to suppress some reflexes by conscious action. This can be demonstrated by trying to prevent yourself blinking when air is blown lightly into your eye. Some individuals can suppress the reflex and keep their eyes open.

## ⚙ LET'S THINK ABOUT THIS

Answer the following questions.

**1** Describe the effect of sympathetic stimulation on (**i**) sweating (**ii**) heart rate (**iii**) bronchi and (**iv**) the digestive system.

**2** Which line in the table below correctly describes the neural pathway shown?

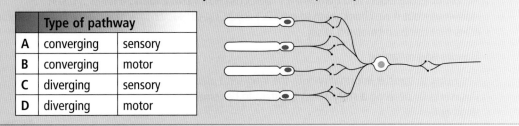

|   | Type of pathway | |
|---|---|---|
| A | converging | sensory |
| B | converging | motor |
| C | diverging | sensory |
| D | diverging | motor |

# BEHAVIOUR I

## MEMORY

Memory involves all the processes that enable us to store, retain and retrieve information. Memories include information from the senses, as well as the emotions that we experience.

To form new memories, information coming into the brain is changed into a usable form (**encoded**) and **stored**. There must also be a method for **retrieval**, so that we can access the information at a later date.

### Encoding

During **encoding**, information that we see, hear, think and feel is changed into a form that can be stored as a memory. Encoding can be done by use of different codes such as:

- an **acoustic code** – a sound image is created
- a **visual code** – a visual image is created.

### Storage

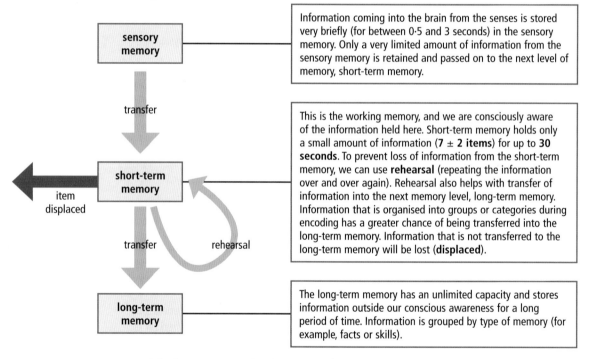

| | |
|---|---|
| **sensory memory** | Information coming into the brain from the senses is stored very briefly (for between 0·5 and 3 seconds) in the sensory memory. Only a very limited amount of information from the sensory memory is retained and passed on to the next level of memory, short-term memory. |
| transfer | |
| **short-term memory** (item displaced) | This is the working memory, and we are consciously aware of the information held here. Short-term memory holds only a small amount of information (**7 ± 2 items**) for up to **30 seconds**. To prevent loss of information from the short-term memory, we can use **rehearsal** (repeating the information over and over again). Rehearsal also helps with transfer of information into the next memory level, long-term memory. Information that is organised into groups or categories during encoding has a greater chance of being transferred into the long-term memory. Information that is not transferred to the long-term memory will be lost (**displaced**). |
| transfer / rehearsal | |
| **long-term memory** | The long-term memory has an unlimited capacity and stores information outside our conscious awareness for a long period of time. Information is grouped by type of memory (for example, facts or skills). |

Areas of the brain associated with storage of memories include the **limbic system**. This lies deep in the brain, extending into the temporal lobe of the cerebrum.

### Methods to enhance storage of memories

#### Chunking

The short-term memory span is $7 \pm 2$ items. However, we can boost memory by putting related information into groups (**chunks**), forming larger meaningful items that can be stored.

For example, the letters and numbers ITV1BBCCH4CH5ITV2MTV form 20 items if taken individually, but if we chunk them to get ITV1-BBC-CH4-CH5-ITV2-MTV they form six meaningful items that can be memorised easily.

#### Elaborating meaning

It is easier to transfer information into the long-term memory if we elaborate on its meaning. This works by devising a story about the item to be remembered – making the information stand out.

contd

## MEMORY contd

### Retrieval

It is easier to retrieve information if you are in the same setting or context as you were when the information was encoded. So, particular sights, sounds, smells or emotions act as **contextual cues**, triggering retrieval of a memory. For example, visiting your old home or school can evoke memories of your childhood, or the smell of a perfume can evoke memories of a particular person or an event.

### A molecular basis for memory

Two molecules are thought to have a role in the storage of memories: **acetylcholine** and **NMDA**.

#### Acetylcholine

The neurotransmitter **acetylcholine** is prevalent in part of the limbic system called the **hippocampus**. The number of acetylcholine-producing cells is decreased in patients suffering from memory loss in **Alzheimer's disease**, indicating that acetylcholine is involved with connecting neurones during the formation of memories.

#### NMDA

Nerve cells in the **hippocampus** of the limbic system have a high concentration of a receptor molecule called **NMDA** receptor.

> ### DON'T FORGET
>
> Contextual cues are built up during encoding; elaboration adds to the number of contextual cues for each piece of information. The more contextual cues present, the easier the recall.

## ⚙ LET'S THINK ABOUT THIS

You should be familiar with the method used to determine the capacity of the short-term memory.

1 Subjects are asked to put pens/pencils down before starting the experiment (to prevent cheating).
2 A series of three letters or numbers is read out, one at a time, in a monotone voice and at regular speed.
3 Students pick up pens/pencils and write down the series.
4 The procedure is repeated with the number of items in the series increasing each time.
5 A second set of letters or numbers is then used to increase the test's reliability.
6 The maximum number of items that can be recalled correctly by each subject is taken as their memory span, with class results being pooled to determine the maximum and minimum memory span for the class.

A typical set of results would show the memory span ranging from 5 to 9 items, that is $7 \pm 2$.

You should also be familiar with the method used to investigate the **serial position effect**.

1 Subjects are asked to put pens/pencils down before viewing the items (to prevent cheating).
2 Twenty items are shown to the subject, one at a time, for 5 seconds (limiting rehearsal time).
3 After all twenty items have been shown, subjects pick up pens/pencils and write down as many items as they can recall.
4 The experiment is repeated with a fresh set of items (to increase reliability).

The graph opposite shows a typical set of results for this experiment.

Why is this pattern of results obtained?

Items at the beginning of the sequence are remembered well because there has been time for rehearsal and transfer to the long-term memory.

Items at the end of the sequence have not been displaced from the short-term memory and are, therefore, also recalled well.

However, items in the middle of the sequence are poorly recalled as they have been displaced from the short-term memory.

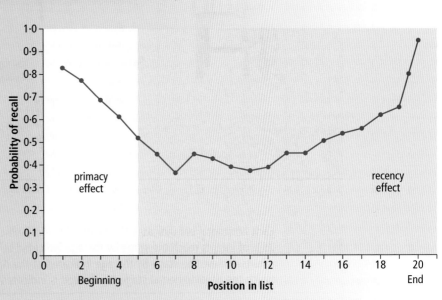

# BEHAVIOUR II

Human development, including behaviour, is influenced by three factors: **maturation**, **inheritance** and the **environment**.

## MATURATION

**Maturation** is the sequence of developmental stages through which all humans pass as they grow older. The sequence of stages through which a child passes is similar in all cultures, indicating that the sequence is inherited. However, the rate at which a child progresses through the sequence is affected by the both genetic and environmental factors (such as nutrition and levels of encouragement).

### Development of walking

**Motor development**, as demonstrated by learning to walk, is largely influenced by maturation. The sequence of stages through which infants pass (shown in the table below) depends partly on adequate muscle and bone development to support the infant's weight, and partly on the level of **myelination** in the nervous system. Formation of myelin sheaths increases the speed of nerve impulses, allowing the infant to gain control over skeletal muscles. As myelination is completed in the upper limbs before the lower limbs, infants gain control of their arms before their legs.

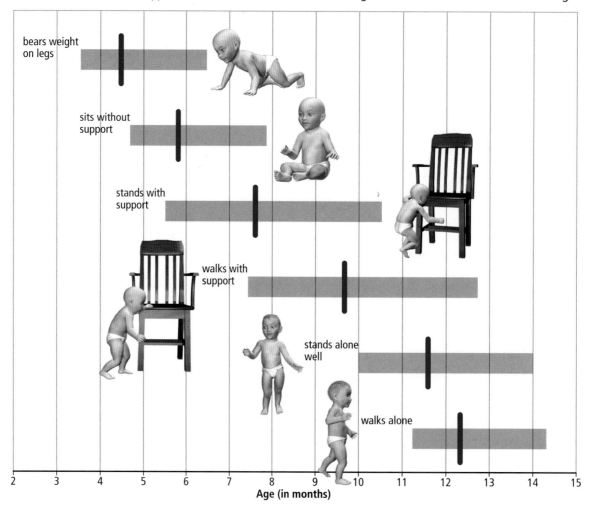

bears weight on legs

sits without support

stands with support

walks with support

stands alone well

walks alone

Age (in months)

Each horizontal bar indicates the range of ages at which each stage of development is reached. 25% of infants reach the stage by the age indicated at the left end of the bar, and 90% of infants achieve the stage by the age indicated at the right end of the bar. The average age at which the developmental stage is achieved is indicated by the vertical bars.

## INHERITANCE

The genes that are present in any individual are determined at fertilisation, when sperm and egg fuse. Sometimes an inherited condition alters the development of the nervous system and can affect the individual's behaviour. You should be familiar with two such inherited conditions.

### PKU

**Phenylketonuria** (PKU) results from a defect in a single recessive gene. The enzyme which breaks down phenylalanine is not produced, allowing phenylalanine to build up in the body to toxic levels, causing abnormal brain development and mental disability (see page 9). If untreated, affected individuals usually die before reaching 30 years of age.

### Huntington's chorea

**Huntington's chorea** is caused by a single dominant gene. Affected individuals show no symptoms until after 30 years of age, when degeneration of areas of the brain causes a progressive deterioration in the individual's motor and mental function.

> **DON'T FORGET**
>
> Inheritance influences our ability to speak. Maturation determines the stages through which we progress. The environment influences the language that we speak.

## ENVIRONMENT

The environment that an individual experiences as they develop interacts with the inherited characteristics to determine development and behaviour. The environmental factors that influence development depend on **social group**, **culture** and the **family** to which the individual belongs.

### Twin studies

Traditionally, the most popular way to investigate the interaction of inheritance and environmental factors in development is to study intelligence quotient (**IQ**) in twins. Intelligence is an umbrella term covering a range of skills, such as the ability to learn and adapt. Remember, however, that IQ tests only examine a very narrow range of skills.

**Monozygotic twins** are genetically identical, so any differences between them must arise through the influence of environmental factors. Results show that monozygotic twins who have been raised in the same family have IQs that are more closely matched (that is have a **higher correlation**) than monozygotic twins who have been raised by different families. This difference must be due to environmental factors, since the twins are genetically identical.

## LET'S THINK ABOUT THIS

1 Put the following pairs of individuals in order, with the pair whose IQs are likely to correlate the most at the top.
   **(a)** monozygotic twins reared apart
   **(b)** dizygotic twins reared together
   **(c)** monozygotic twins reared together
   **(d)** dizygotic twins reared apart.

2 A child born to Scottish parents but raised in India may speak English but with an Indian accent. Explain the influence of maturation, inheritance and environment on the development of speech in this individual.

# BEHAVIOUR III

## INFANT ATTACHMENT

In humans, there is a long period of dependency, when children develop social and communication skills that allow them to function successfully throughout life. A strong emotional bond (**infant attachment**) develops between a child and the primary carers, providing the child with a secure base from which to explore the surroundings. Children who form secure attachments in infancy are thought to develop into more resilient and trusting adults.

## COMMUNICATION

Humans display complex behaviour, partly due to our highly developed language skills. Language is supported by **non-verbal communication**, a series of actions that obey **social** and **cultural rules**.

### Language

**Language** involves the use of words (spoken or written) to represent information. Through this representation, we organise thoughts and communicate them to others, potentially speeding up our rate of learning.

### Non-verbal communication

When we interact with each other, wordless signals are passed between us. These indicate attitudes and emotions, and add to any verbal message that they accompany. **Non-verbal communication** is, therefore, important in forming and maintaining all successful relationships, including the relationship between an infant and its carers. Initially, this relationship must be based on non-verbal communication, as the infant's language skills have yet to develop (think of the effect that an infant's smile has on the mother and father).

Non-verbal communication can either **reinforce** or **contradict** spoken language. If the non-verbal message is in agreement with the verbal message, it helps to make the message clear. For example, banging your fist on a table when shouting at someone reinforces anger; smiling while praising someone reinforces the praise.

However, if the non-verbal and verbal messages are not in agreement, the message becomes confused. For example, by fidgeting and avoiding eye contact while telling the truth, our body language implies that we are lying. We may deliberately use contradictory non-verbal and verbal messages, either to unnerve someone or in humour when teasing.

contd

## COMMUNICATION contd

The table below shows some forms of non-verbal communication.

| Type of non-verbal communication | Description |
|---|---|
| personal space | The physical distance that is maintained between two people indicates the level of intimacy between them. Close friends and family may be allowed to 'invade our personal space', but we generally keep others at a distance. |
| eye contact | We use eye contact (both frequency and length of gaze) to signal our emotions, to define our status and to regulate interactions between us. For example:<br><br>● if you are attracted to an individual, you will repeatedly try to catch their gaze<br>● we can indicate anger by staring continuously at someone<br>● subservient individuals keep their eyes down, avoiding eye contact with someone whom they consider to be superior. |
| facial expression | We use a vast range of facial expressions to convey emotions, such as:<br><br>● frowning – disapproval<br>● smiling – interest or attraction<br>● winking – humour. |
| gestures | Some gestures can substitute for verbal language (a nod can replace a verbal 'yes'; waving signals 'hello' in many cultures). Other gestures indicate emotions (for example, a clenched fist can indicate anger). |
| body posture | We convey our attitude through the posture that we maintain. For example:<br><br>● students convey that they are paying attention by sitting upright and not slouching<br>● we convey interest in another individual by angling our bodies towards them<br>● we show that we are relaxed by sitting back with arms unfolded and legs extended.<br><br>Friendship can be conveyed by mimicking an individual's body posture. |
| tone, volume and pitch of voice | Non-verbal aspects of speech add to the emotions conveyed by the verbal message. For example:<br><br>● a low volume can convey nervousness<br>● a high pitch can signal excitement. |

### LET'S THINK ABOUT THIS

Answer the following questions.

1 Explain why there is a long period of dependency in humans.

2 Non-verbal communication is important in infancy.
   (i) Explain why.
   (ii) Give two examples of non-verbal communication that takes place between a mother and her infant.

# BEHAVIOUR IV

## EFFECT OF EXPERIENCE

Human behaviour is modified in the light of experience. Experience is gained in several ways.

### Practice

The repeated use (practice) of **motor skills** causes an increased number of synapses to form in the brain, producing a **motor pathway**. Skills such as writing or driving a car develop through the formation of this **motor memory**.

### Imitation

In many situations, the preferred method of learning is through imitation. Here, a behaviour is observed and then copied: for example, children may learn to cross the road by copying adults. Imitation is often used during **training**.

### Trial and error

When an individual is **rewarded** for performing a particular behaviour (for example, a parent smiling and interacting with an infant when it says a new word), the behaviour is **reinforced** and is likely to be repeated. When the behaviour is not rewarded, it may become **extinct**.

Individuals can be trained to display positive behaviour through **shaping**. Here, each time performance moves closer to the desired behaviour, a reward is given, until – eventually – the desired behaviour is displayed. This technique can be used, for example, children are toilet-trained.

### Generalisation and discrimination

Sometimes an individual responds in a similar manner to different but related stimuli (for example, a fear of *all* spiders). This indiscriminate response is called **generalisation**. However, when the individual is able to give a different reaction to related stimuli (they are scared of big spiders, but not small ones), the response is termed **discrimination**.

## GROUP BEHAVIOUR AND SOCIAL INFLUENCE

An individual's emotions and behaviour are influenced by the presence of others.

### Identification

Individuals alter their behaviour to be more like someone they admire in a process called identification. This role model may be a family member, friend or colleague, a celebrity or sports star. Identification is used by advertisers who employ celebrities to sell their products.

### Deindividuation

Deindividuation involves the loss of personal identity in a group situation. Having become anonymous within the group, individuals lose their sense of judgement and behave in a less appropriate manner (as seen in the case of anti-social behaviour displayed by football hooligans).

### Internalisation

Internalisation is a change in an individual's beliefs due to persuasion. Governments use persuasion in their health campaigns.

### Social facilitation

Individuals often perform better in tasks when they are in the company of others (either other competitors or an audience). For example, individuals often run faster when competing against others in a race than when they run alone.

## LET'S THINK ABOUT THIS

A finger maze can be used to construct a learning curve (shown below), either by measuring the time taken to complete the task, or by counting the number of errors made when completing the task.

The graph shows that, as the number of trials increases (and the subject practises the task), performance improves until the task cannot be carried out any faster.

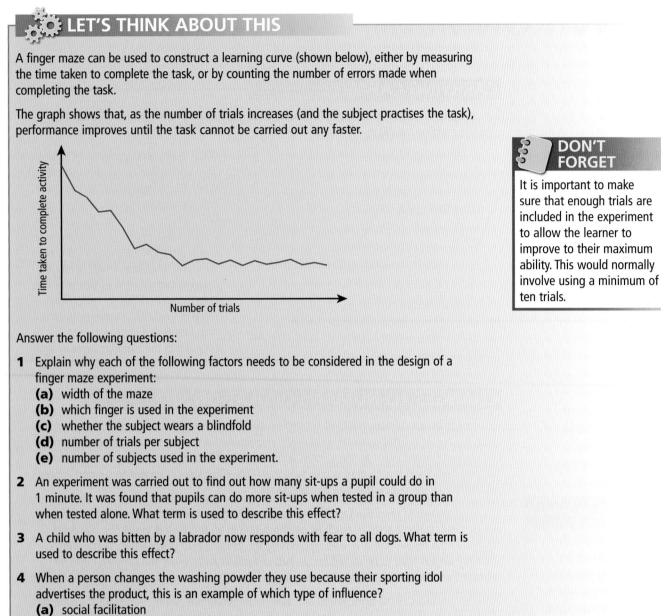

**DON'T FORGET**

It is important to make sure that enough trials are included in the experiment to allow the learner to improve to their maximum ability. This would normally involve using a minimum of ten trials.

Answer the following questions:

1   Explain why each of the following factors needs to be considered in the design of a finger maze experiment:
    **(a)** width of the maze
    **(b)** which finger is used in the experiment
    **(c)** whether the subject wears a blindfold
    **(d)** number of trials per subject
    **(e)** number of subjects used in the experiment.

2   An experiment was carried out to find out how many sit-ups a pupil could do in 1 minute. It was found that pupils can do more sit-ups when tested in a group than when tested alone. What term is used to describe this effect?

3   A child who was bitten by a labrador now responds with fear to all dogs. What term is used to describe this effect?

4   When a person changes the washing powder they use because their sporting idol advertises the product, this is an example of which type of influence?
    **(a)** social facilitation
    **(b)** internalisation
    **(c)** deindividuation
    **(d)** identification

# ENVIRONMENT I

## CARRYING CAPACITY

The **carrying capacity** is the maximum number of individuals of one species that can be supported within a habitat.

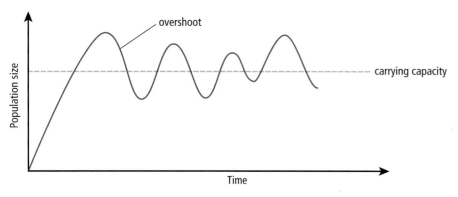

Often, a population increases in size until it overshoots the carrying capacity. When this happens, the environment can no longer support the population; lack of food, overcrowding, predation and increased incidence of disease may cause the population size to decrease. When the population falls below the carrying capacity, the environment is able to recover. This, in turn, allows more individuals to survive, and the population size increases until it overshoots the carrying capacity again.

## CHANGES TO HUMAN POPULATION SIZE

For most of human history, population size remained small and relatively stable. Early humans lived as **hunter-gatherers**, with the population harvesting natural products from the surrounding habitat (such as fruits, roots and honey) and hunting or trapping animals. The carrying capacity in this form of economy is low – the ecosystem is unable to support a dense population, so the population size remains small.

The development of primitive farming methods, cultivation of plants (such as wheat and rice) and domestication of animals (such as pigs) resulted in an **increased yield**. The nomadic lifestyle of the hunter-gatherer was replaced by the formation of **permanent settlements**, allowing surplus food to be stored for periods of famine. As a result, the carrying capacity increased and human population size began to rise.

With the advances of the agricultural and industrial revolutions having improved the food supply, given better medical care and enhanced sanitation, the death rate (particularly of infants and children) has decreased dramatically, causing exponential growth of the human population. How long our species' population can continue to grow at this rate is unknown.

> **DON'T FORGET**
>
> In exponential growth, the growth rate is proportional to the size of the population; that is, the larger the population size, the faster the population grows.

## POPULATION PYRAMIDS

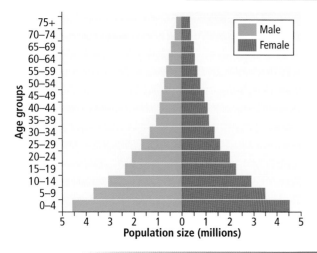

The study of **demographic trends** focuses on human population change. Most countries carry out a ten-yearly census of their population. From this data, **population pyramids** can be constructed and used by governments to forecast changes in the age structure of their population, allowing planning for future **housing**, **education** and **health-care** needs to take place.

### Developing countries

The population pyramid of a developing country is typically triangular in shape, with a wide bottom and a narrow top. The population has a:

● high birth rate – due to lack of access to birth control and education
● high death rate – due to poor nutrition and medical care
● low life expectancy.

contd

## POPULATION PYRAMIDS contd

A developed country has a more stable population pyramid, with a broader top and narrower bottom compared to that of a developing country. The population has a:

- low birth rate – due to access to birth control and education
- low death rate – due to good medical care, education and nutrition
- high life expectancy.

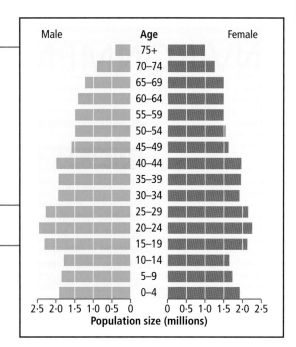

# FACTORS AFFECTING EXPONENTIAL GROWTH

## Overcoming predation

The manufacture of weapons has allowed humans to protect themselves from predators.

## Increased food availability

Modern agricultural methods (including use of fertilisers, herbicides and pesticides), together with modern animal husbandry and fishing techniques, and the application of technology, have resulted in increased food production.

## Reduced child mortality

Child mortality has decreased as a result of improved medical care (for example vaccination programmes), hygiene, sanitation and nutrition. More individuals, therefore, survive to reach reproductive age and produce offspring of their own.

## Increased life expectancy

Improved nutrition, sanitation and medical care have resulted in individuals living longer.

## Female fecundity

Female fertility (fecundity) has increased through:

- increased life expectancy
- the earlier onset of puberty.

## Cultural habits

Suckling an infant results in the production of hormones that inhibit ovulation, making the female less fertile. However, the length of time over which women breastfeed their children has decreased, resulting in successive pregnancies being closer together.

# CONTROL OF BIRTH RATE

Development of methods of birth control, such as male and female sterilisation, contraceptives and abortion, has caused the birth rate to be reduced. This is most obvious in developed countries where lifestyle and economic choices have made small family size more desirable (raising a child takes up a lot of time and is expensive). In some countries, however, large families are culturally desirable or may provide a larger workforce, giving families greater financial security.

# LET'S THINK ABOUT THIS

Population size in developing countries has the greatest capacity for increase. Explain why.

# ENVIRONMENT II

Population-limiting factors include food supply, water supply and disease.

## FOOD AS A LIMITING FACTOR

After the agricultural revolution, the development of new farming methods allowed food production to increase. At present, on a global scale, more food is produced than is required. However, food distribution on the planet is not equal, with **famine** and **malnutrition** affecting populations in some areas (mainly developing countries). Malnutrition results when a diet is deficient in one or more essential nutrients, and is one of the main causes of death among children under 5 years old. Poor nutrition not only reduces life expectancy but also affects reproductive health, with female fertility being reduced and puberty being delayed.

Altered land use, application of chemicals and new breeding methods have all contributed to increased food production as follows.

### Land use

Land use has been altered to allow for increased food production. Ecosystems have been affected through these alterations.

- **Deforestation** – habitat loss results in loss of species from the area, topsoil may be washed away resulting in poor soil quality, and the water cycle is disrupted.
- **Monocultures** – large areas of land are cleared and used to produce one crop. Natural succession is prevented, limiting the number of species in the ecosystem.
- **Cash crops** – crops such as cocoa and coffee are grown to provide income for the country. However, while not growing food crops for the population, the risk of starvation is increased.

### Introduction of chemicals to the ecosystem

Food production has been massively increased by the use of:

- **fertilisers** – replace nutrients removed from the soil by successive harvests
- **herbicides** – kill weeds that would be in competition with the crop species
- **pesticides** – kill insect pests that reduce crop yield
- **fungicides** – prevent growth of fungal spores on the crops.

Each of these chemicals can have a detrimental effect on the ecosystem, and their use must be controlled. Excessive use of fertilisers can cause a build-up of nutrients in water ecosystems, resulting in **algal blooms**. Pesticides, herbicides and fungicides may be toxic and can enter the food chain.

### Breeding methods

**Selective breeding** is used to increase the yield and quality of crop and animal products. The development of **genetic engineering** (where genes for a desired characteristic from one species can be placed in the chromosomes of a different species) has allowed new characteristics (such as drought resistance in plants) to be introduced into crop and animal species.

## WATER AS A LIMITING FACTOR

Only 2·5% of all water on Earth is fresh water, with less than 0·1% of this being freely available. The increasing size of the human population has put a huge demand on this supply, with uses including **domestic**, **agricultural** and **industrial**.

### Domestic

Increased population size and improving standards of living have resulted in an increased demand for fresh water. Some water is used for drinking and cooking, but most is used for sanitation and bathing.

contd

## WATER AS A LIMITING FACTOR contd

### Agricultural

- Two thirds of all available water is used for agricultural purposes, including irrigation of crops and provision for livestock.
- Cultivation of marginal land at the edge of deserts increases the demand on a very limited water supply. It can also result in a decrease in soil quality (**desertification**) due to overgrazing or overcultivation of crops.
- Deforestation to produce agricultural land disrupts the water cycle. For example, there is increased water run-off from the land, and reduced evaporation may result in decreased rainfall in the area.

### Industrial

Water is in high demand in industry where it is involved in many processes, for example cooling.

To meet the demand on fresh water caused by the population increase, water retrieval and storage systems have been developed, including **wells**, **reservoirs** and **dams**.

> **DON'T FORGET**
> Climate change has resulted in altered patterns of rainfall, with floods and droughts occurring more frequently.

## DISEASE AS A LIMITING FACTOR

Disease became the most important limiting factor on population growth after agricultural methods and urbanisation stabilised food supplies. **Epidemics** spread through densely populated areas, and increased levels of migration carried the disease from town to town. In modern times, control of disease is achieved in various ways.

### Vaccination programmes

The development of **vaccines** has had an important effect on reducing the number of deaths due to disease. The **World Health Organisation** (WHO) set up a vaccination programme against smallpox, causing it to be completely eradicated; similar programmes aim to eradicate other childhood diseases such **polio** and **measles**. However, progress is hampered by poor medical infrastructure in many developing countries, where the death rate remains high.

### Improved sanitation

Where **sewage** and **drinking water** are kept separate, the incidence of waterborne diseases such as **cholera** and **typhoid** is greatly decreased. However, in many developing countries there is no access to clean drinking water.

### Improved hygiene

Improved **personal hygiene** (such as hand-washing and bathing) and **medical hygiene** (for example, use of disinfectants and sterile surgical techniques) have reduced the spread of disease.

## LET'S THINK ABOUT THIS

Most deaths in developed countries are the result of cancer and cardiovascular diseases, but in developing countries result from poor health care, nutrition and lack of sanitation. What factors have contributed to the higher incidence of cancer and cardiovascular disease in developed countries?

Make up a bullet-point list of advantages and disadvantages of the use of chemical fertilisers in agriculture.

# ENVIRONMENT III

## DISRUPTION OF FOOD CHAINS

Established ecosystems, such as woodlands and lochs, contain a large variety of species (high levels of **biodiversity**) and complex food webs. Complex food webs are more stable than simple food webs because they have many interconnections between species. If one species is lost from the web, there is usually an alternative source of food. In a simple food web, however, loss of one species can lead to the loss of several other species that have no alternative food source. Use of land for agricultural purposes, including the application of herbicides and pesticides, and the formation of monocultures, results in a loss of complexity of food webs. In addition, development of land for housing and industry has caused loss and fragmentation of habitats, and destruction of transport corridors through which species move, leading to species isolation and loss.

## EFFECT OF CHEMICALS

Chemicals may be introduced into ecosystems through accidental spillage (for example, oil spills) or from agricultural or other uses (for example, fertilisers). You should be familiar with the effects shown below.

### Resistance to pesticides

The use of pesticides has led some insects and other pests to develop a resistance to these chemicals.

### Build-up of chemicals in the food chain

When chemicals don't break down, they may enter the food chain and become more concentrated. For example, **DDT** is an insecticide that enters the food chain and has a catastrophic effect on non-target species.

### Run-off

Chemicals sprayed onto agricultural land (such as fertilisers containing nitrates and phosphates) are washed into and pollute waterways. Where nitrates and phosphates build up in waterways, the number of algae can multiply rapidly to form an algal bloom (see page 87).

## DISRUPTION OF THE NITROGEN CYCLE

Nitrogen is an essential component of DNA, RNA and protein. It is recycled in the nitrogen cycle by way of five main processes that allow transfer of nitrogen between molecules.

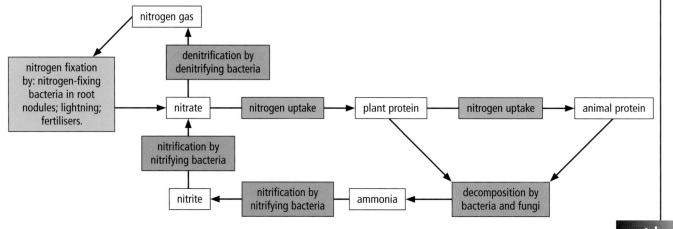

contd

## DISRUPTION OF THE NITROGEN CYCLE contd

Disruption of the nitrogen cycle by humans has the following effects.

### Fertiliser run-off

The concentration of nitrites and nitrates in rivers increases when too much fertiliser is added to agricultural land, running off into rivers. This causes algae in the water to multiply, producing an **algal bloom**. When the algae die, bacteria feed on the cells and multiply. The increase in bacterial respiration causes the oxygen concentration in the river water to decrease. As most aquatic organisms cannot survive in low oxygen concentrations, there is a decrease in the number of species in the ecosystem. This can be prevented by applying the recommended concentration of fertiliser to farmland.

### Inadequate sewage treatment

Inadequate sewage treatment results in raw sewage entering rivers and seas, again causing the nitrite and nitrate levels to increase, risking the production of algal blooms. Bacteria in the water multiply, feeding on the raw sewage and the increased quantity of decaying plant material. The rate of bacterial respiration increases and the oxygen concentration in the water decreases, resulting in the death of fish and other aquatic animals. With an increasing human population, it is important that efficient sewage treatment is carried out, ensuring that no raw sewage is released.

### Contamination of drinking water

While nitrates in our drinking water are not toxic to adults, they can have a more serious effect on infants. Within the gut, nitrates are converted to nitrites, which can bind to haemoglobin, interfering with its ability to transport oxygen around the body.

http://bcs.whfreeman.com/thelifewire/content/chp58/5802004.html

**DON'T FORGET**

The more bacteria in water, the lower the oxygen concentration.

## LET'S THINK ABOUT THIS

High nitrate levels in rivers can cause algal blooms.

1   Explain how the nitrate level in rivers may increase.

2   Explain the effect that algal blooms can have on rivers.

3   River water was sampled from above and below a sewage outlet. Which of the following comparisons would be correct?

|   | Water taken from above sewage outlet | Water taken from below sewage outlet |
|---|---|---|
| A | low oxygen concentration | few bacteria |
| B | low oxygen concentration | many bacteria |
| C | high oxygen concentration | few bacteria |
| D | high oxygen concentration | many bacteria |

# ENVIRONMENT IV

## DISRUPTION OF THE CARBON CYCLE

Carbon is an essential component of all organic molecules and is recycled through the processes shown below.

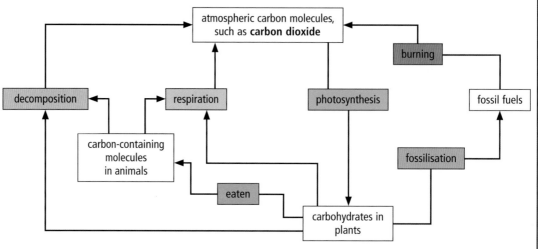

The carbon cycle has been disrupted by humans in the following ways.

| Process | Disruption by humans |
|---|---|
| carbon uptake through photosynthesis | • The increasing need for land for farming and housing has resulted in large areas of deforestation. As a result, the number of trees that is present to absorb **carbon dioxide** from the atmosphere has decreased.<br>• Overgrazing has also reduced pasture to desert, adding to the decrease in photosynthesis. |
| decomposition | • Increasing numbers of farmed cattle and other livestock are responsible for adding **greenhouse gases,** such as carbon dioxide and methane, to the atmosphere.<br>• As oxygen is absent, **methane gas** is released from decomposing animal and plant remains that have been dumped in landfill sites. Landfill gases can be collected and used as an energy source, lessening their impact on the atmosphere.<br>• Methane is produced from rice paddies through the anaerobic decomposition of organic matter in the soil. |
| burning | Burning fuels (such as fossil fuels and wood) releases vast quantities of **carbon dioxide** into the atmosphere. |

http://epa.gov/climatechange/kids/carbon_cycle_version2.html

## GLOBAL WARMING

**Greenhouse gases** such as **carbon dioxide** and **methane gas** act like a blanket around the planet, reducing heat loss and preventing the Earth from becoming completely frozen. However, release of these gases through human impact on the environment (there is now about 30% more $CO_2$ in the atmosphere than there was 150 years ago) has resulted in rising global temperatures and **climate change**.

### Effects of global warming

#### Changes to weather patterns

Changes to rainfall patterns are becoming more common, with some areas being exposed to periods of either heavier rainfall or drought. The number and strength of tropical storms has also increased, with the El Niño effect causing changes to ocean currents. The number of Category 4 and 5 storms that has been recorded over recent decades has greatly increased. For example, Hurricane Katrina in 2005 was one of the most dangerous storms ever to hit the USA.

#### Increased flooding

As global temperatures increase and polar ice caps melt, sea levels will rise, causing increased flooding and permanent loss of low-lying land areas. This could have a serious impact, as it will affect much of the land that we inhabit and use for agricultural purposes.

#### Ecosystem shifts

Increasing temperatures and changes to moisture patterns will disrupt ecosystems, with species either becoming extinct or colonising new areas if they cannot adapt. A recent survey has identified a shift in the range of many species, with the area colonised drifting slowly towards the poles. A similar move in distribution has been found in alpine ecosystems, where a vertical shift in distribution has taken place.

**DON'T FORGET**

Carbon dioxide and methane gas contribute most to global warming.

## LET'S THINK ABOUT THIS

The changing climate in the UK is affecting the behaviour of our butterfly species. Not only are species such as the Essex Skipper extending their range, with many being found further north, but they are also appearing earlier in the year. Warmer temperatures have also extended the breeding period, enabling some species to reproduce several times in the summer months.

Use the internet to research changes in the behaviour pattern of other species in response to climate change.

Which of the following processes would result in an increase in atmospheric carbon dioxide?

**(a)** photosynthesis
**(b)** decomposition
**(c)** nitrogen fixation
**(d)** fossilisation.

# EXTENDED-RESPONSE QUESTIONS

## ESSAY GUIDE

When writing essays for the Higher Human Biology exam, there are some easy-to-follow guidelines that will help you to maximise your grade.

- For each essay question, you have the choice of two titles. Don't jump at the essay that you think looks easier. Make a very brief plan for each title to make sure you pick the one that will give you more marks.

- Use your essay plan to ensure that you stick to the question being asked in the title. You can waste an awful lot of time writing down irrelevant information that will not gain marks.

- Human Biology essays should be an organised list of facts, placed in the correct context. So, don't waste time writing an opening paragraph (as you would in English); get straight to the point.

- In Essay 1, the marks allocated to each section of the essay are given on the paper. These marks are purely for knowledge. It is a good idea to write the facts in an ordered bullet-point list under the subheadings given in the essay title.

- Essay 2 has eight marks allocated for knowledge. An additional one mark is awarded for relevance and one mark for coherence (flow). If you are short of time, it may be wise to bullet-point this essay and not worry about the coherence mark.

- Spelling and grammar are not being marked. As a general rule, if you spell the word as it would be said, so that it cannot be mistaken for any other biological term, the mark will be awarded.

- Diagrams are a very useful way of displaying a lot of information, but they can take time to draw. If you are using diagrams, they should be quick sketches and must be labelled. If you are adding arrows to diagrams, they should point in the right direction and must have arrowheads. Do not colour in diagrams, as this wastes time and will gain no extra marks.

- Finally, do not score out your essay plans. The examiner will look for marks in everything you have written, and any facts that you had placed in your plan but missed in your essay will be credited.

## EXAMPLES OF EXTENDED-RESPONSE ESSAY ANSWERS

### Essay 1

Give an account of homeostasis under the following headings:

**(a)** Control of water balance (5 marks)

**(b)** Hormonal control of blood glucose level (5 marks)

#### Control of water balance

- Osmoreceptor cells in the hypothalamus detect changes in water concentration in the blood.
- Nerve impulses send information from the hypothalamus to the pituitary gland.
- If the water concentration in the blood is too low, the pituitary gland produces more antidiuretic hormone (ADH).
- If the water concentration in the blood is too high, the pituitary gland produces less ADH.

contd

## EXAMPLES OF EXTENDED-RESPONSE ESSAY ANSWERS contd

- ADH travels in the blood to the nephrons in the kidney.
- Increased levels of ADH cause the ascending limb of the loop of Henlé, distal convoluted tubule and collecting duct to be more permeable to water.
- This increases water reabsorption, and a low volume of concentrated urine is produced.

### Hormonal control of blood glucose concentration

- After a meal, the blood glucose level increases.
- Receptor cells in the pancreas detect blood glucose concentrations.
- High blood glucose levels cause the pancreas to secrete the hormone insulin.
- Insulin travels in the blood to the liver and causes liver cells to convert glucose into glycogen.
- This brings the blood glucose level back down to normal again.
- Between meals, the blood glucose level decreases.
- Receptor cells in the pancreas detect this decrease and produce the hormone glucagon.
- Glucagon passes in the blood to the liver and causes liver cells to convert glycogen to glucose.
- This increases the blood glucose level again.

## Essay 2

Give an account of the homeostatic control of water balance and glucose concentration.

Negative feedback maintains the optimum levels of water and glucose in the body. Blood water concentration is detected by osmoreceptors in the hypothalamus. The hypothalamus sends nerve impulses to the pituitary gland. When there is a low water concentration in the blood, the pituitary gland is stimulated to produce more antidiuretic hormone (ADH). ADH passes in the blood to the kidneys, where it acts on cells of the ascending limb of the loop of Henlé, distal convoluted tubule and collecting duct, making the cells more permeable to water. This causes more water to be reabsorbed, and a low volume of concentrated urine is produced. The water concentration of the blood is returned to normal. When the water concentration of blood goes too high, less ADH is produced, making the kidney tubule less permeable to water. A high volume of dilute urine would then be produced.

Glucose concentration is detected by cells of the pancreas. After a meal, the blood glucose level increases above normal, and the pancreas is stimulated to produce the hormone insulin. Insulin travels in the blood to the liver, where it causes liver cells to convert glucose into glycogen. The blood glucose level now returns to normal. When the blood glucose level decreases below normal, as it does between meals, the pancreas is stimulated to produce the hormone glucagon. Glucagon travels in the blood to the liver. Liver cells now convert glycogen into glucose, which is released into the blood, raising the blood glucose level back to normal.

# PROBLEM-SOLVING

## DEVELOPING THE RIGHT SKILLS

Problem-solving involves assessing and processing information that has been given to you to interpret. By mastering these skills, you will be maximising your chances of a good pass in the exam. You will benefit from using past papers to practise problem-solving questions, enhancing your technique and identifying areas that require further revision. The advice below covers the key problem-solving skills as defined in the Higher Human Biology arrangements.

## SELECTING RELEVANT INFORMATION

- Often, you will be asked to select information from a graph. Read the x- and y-axes carefully, taking note of the variables and units being used. Use a ruler to help you read values from both axes – be accurate.
- If two sets of values have been plotted with different y-axes, check for differences in scale. Be careful to read the correct y-axis for each line or set of values.
- Make sure that you read the question properly, underlining any relevant information.

## PRESENTING INFORMATION

- It is likely that you will be asked to take information from a table and present it in the form of a graph.
- Use the column headings from the table as the axis labels on your graph. Copy the headings exactly as they have been written, including units.
- The variable that has been deliberately altered in the experiment should be placed on the x-axis. This is usually the set of data in the left column of the table. The variable that has been obtained as the result goes on the y-axis of the graph.
- Your graph must cover 50% or more of the graph paper. Decide on a suitable scale, making sure that you can plot all the points. The scale should be linear – each division must be worth the same number of units; that is, the number of squares between 10 and 20 must be the same as the number of squares between 20 and 30 on the same axis.
- **Bar graphs**: bars should be of equal width, using a ruler to draw a line around the boundary of the bar. You may be asked to include two sets of data and so should distinguish between sets by shading bars and including a key.
- **Line graphs**: mark points with a lightly-drawn X and join points together with a fine line drawn using a ruler. Do not extend the line to the zero point on the x-axis **unless** this value has been included in the table, and make sure that the line does not extend beyond the last point given in the table.
- Don't panic if you make a mistake that you cannot correct easily. The exam paper includes an extra grid at the end of the paper.

## PROCESSING INFORMATION

### Averages

Add up all values and divide by the total number of values.

For example, the average of 4, 6 and 8 is $\frac{4 + 6 + 8}{3} = 6$

### Ratios

To express ratios correctly, write down the raw values in the order that the question suggests. Then try to find a number that divides into both values. The final answer must be in as simple a form as possible, so you may have to try dividing the values by several numbers.

## PROCESSING INFORMATION contd

For example, if the number of seeds produced by plant A was 25 and the number of seeds produced by plant B was 35, the ratio of seeds produced by plant B to seeds produced by plant A would be as follows.

35:25 → divide by 5 → 7:5

There is no number that both 7 and 5 can be divided by, so the answer is 7:5.

### Percentages

To calculate a percentage, the value that is to be expressed as a percentage is divided by the total of all values, and the resulting number is multiplied by 100. For example, if 300 eggs were fertilised but only 50 of them hatched, the percentage that survived would be:

$$\frac{\text{value}}{\text{total}} \times 100 = \frac{50}{300} \times 100 = 16\cdot7\%$$

To calculate the exact number when the value is expressed as a percentage of the whole, take the percentage, divide by 100 and multiply by the total number.

For example, if 60% of seeds germinated out of a total of 200, the number of seeds that germinated was:

$$\frac{\text{percentage}}{100} \times \text{total} = \frac{60}{100} \times 200 = 120$$

### Percentage change

To calculate percentage change (either increase or decrease), work out the difference between the initial and final value, divide by the initial value and then multiply by 100.

For example, if a pulse rate before exercise is 70, rising to 120 after exercise, the percentage increase would be:

$$\frac{\text{difference}}{\text{initial value}} \times 100 = \frac{120 - 70}{120} \times 100 = 41\cdot7\%$$

## PLANNING, DESIGNING AND EVALUATING EXPERIMENTS

- Experiments are repeated to give a range of results. An average is calculated to increase **reliability**.
- Only one variable should be altered in each experiment. All other variables must be kept constant to ensure that the experiment is **valid**. (Do use the word 'valid' in your answer; do not use the word 'fair'.)
- Control experiments should be used to ensure that any change in results can be attributed to the variable that was altered. Suitable controls may be to replace organisms with the same mass or volume of glass beads, to use boiled (inactive) tissue or enzymes, or to replace solutions with the same volume of distilled water.
- When identifying variables that should be kept constant, read the question carefully. Usually, you are required to give variables 'not already mentioned' in the question. Good examples to think about are temperature, mass or volume of substances. (Do use the correct term for the variable in your answer; do not use the word 'amount'.)
- Where initial values were not identical, the percentage change must be used to compare results between groups.

> **DON'T FORGET**
>
> Repeats give **reliability**; variables control **validity**.

## DRAWING CONCLUSIONS AND MAKING PREDICTIONS

- When writing a conclusion, describe the overall trend in results. You must include discussion of both variables in your answer. Make sure you describe completely any observed trend. If a graph shows results increasing and then levelling off or decreasing, you must describe both parts of the graph. The clue is in the number of marks assigned to the question: if it is worth two marks, there must be at least two parts to the graph to describe!
- If asked to predict results from a graph, use a ruler to draw a line that extends the graph on the graph paper.

# ANSWERS

## P7
1 Active site
2 Each enzyme acts on only one substrate.
3 High temperatures break bonds that hold the shape of the molecule.
4 X – enzyme concentration; Y – substrate concentration

## p13

| | DNA | mRNA |
|---|---|---|
| Type of sugar present | deoxyribose | ribose |
| Number of strands of nucleotides in molecule | 2 | 1 |
| Bases present | adenine, cytosine, guanine, thymine | adenine, cytosine, guanine, uracil |
| Where molecule is found in the cell | nucleus | nucleus, cytoplasm |

## p17
(a) mRNA codons:    AUG    CAU    CGG    AGU
tRNA anticodons: UAC    GUA    GCC    UCA
amino acid sequence:
methionine    histidine    arginine    serine
(b) (i) AAG (ii) AUU
(c) UCC and GCC and GCA

## p19
(i) active cells contain many large mitochondria with extensive cristae.
(ii) inactive cells contain few small mitochondria with fewer cristae.

## p25
(i) The nucleic acid is destroyed to prevent replication of viruses.
(ii) The protein coat is retained as it acts as the antigen.

## p29
1 X – phosphate; Y – deoxyribose sugar
2 Enzymes or ATP
3 Each new DNA molecule contains one parent DNA strand.
4 Each daughter cell must contain the correct quantity of DNA for normal cell metabolism.

## p33
1 All rollers
2 All rollers
3 75% rollers; 25% non-rollers

## p35
1 All MN
2 50% AB; 25% A; 25% B
3 All sickle-cell trait
4 25% unaffected; 50% sickle-cell trait; 25% sickle-cell anaemia

## p37
1

| Cross 1 genotypes: | $X^NX^n$, | $X^NX^n$, | $X^NY$, | $X^nY$ |
|---|---|---|---|---|
| Phenotypes: | carrier female | colour-blind female | normal male | colour-blind male |
| Cross 2 genotypes: | $X^NX^N$ | $X^NX^n$ | $X^NY$ | $X^nY$ |
| | normal female | carrier female | normal male | colour-blind male |

2   $X^HX^H$   normal female
    $X^HX^h$   carrier female
    $X^hX^h$   haemophilic female
    $X^HY$   normal male
    $X^hY$   haemophilic male

## p43
(i) Testosterone levels are unaffected because, like other hormones, it is transported in the blood.
(ii) Sperm production continues, but the sperm cannot pass into the urethra.

## p47
To induce labour, oxytocin may be injected.

## p49
1 Colostrum provides maternal antibodies, as the baby cannot make its own for the first few months.
2 Progesterone and oestrogen
3 Using fertility drugs
4 Using testosterone, *in vitro* fertilisation or artificial insemination.

## p57
1 Pulmonary vein
2 Oxygenated blood enters through the hepatic artery.
The hepatic portal vein carries blood rich in products of digestion from the small intestine to the liver.
3 (i) Elastic fibres allow the artery to stretch as blood pulses through.
   (ii) Smooth muscle in arteriole walls contracts to narrow the vessel.
4 Veins have valves to prevent backflow of blood.
5 As blood flows from arterial to venous ends of a capillary, there is a decrease in concentrations of oxygen, water and nutrient molecules such as glucose; and an increase in concentrations of carbon dioxide and waste molecules such as urea.

## p59
Foetal haemoglobin should be placed to the left, above the normal adult haemoglobin line.

## p61
(i) Urea increases, glucose decreases
(ii) Urea increases, glucose increases

## p63

1 Hypothalamus
2 Pituitary gland
3 In the blood
4 More water is reabsorbed; smaller volume of urine produced.
5 The urea concentration increases as water is reabsorbed into the blood.

## p67

(a) Pancreas
(b) X – insulin, Y – glycogen
(c) Liver
(d) It is used up in respiration.

# BEHAVIOUR, POPULATIONS AND ENVIRONMENT

## p69

(i) Nerve impulses reach the axon bulb and stimulate the release of noradrenaline from secretory vesicles.
(ii) It is reabsorbed intact by the pre-synaptic cell.

## p73

1 (i) Increases
  (ii) Increases
  (iii) Dilate
  (iv) Decreases peristalsis
2 A

## p77

1 CABD
2 The ability to speak is inherited, the sequence of stages through which the child progresses is determined by maturation and the language spoken is due to the environment. The child speaks English with an Indian accent as this is the form of the language that the child is exposed to.

## p79

1 A long period of dependency allows the development of social and communication skills.
2 (i) The infant cannot communicate verbally, so relies on non-verbal communication.
  (ii) Clapping, smiling

## p81

1 (a) To prevent cheating by using more than one finger.
  (b) To increase validity by controlling variables.
  (c) To prevent cheating by looking at maze.
  (d) Enough trials must be included to allow learning to take place.
  (e) Increases reliability
2 Social facilitation
3 generalisation
4 D

## p83

In developing countries, the high death rate limits population size. With better medication and nutrition, the death rate would decrease and population size could increase.

## p87

1 Nitrate levels increase through the release of raw sewage and from fertiliser run-off into rivers.
2 Decaying material from algal blooms provides food for bacteria. Increased bacteria reduces oxygen levels of river water and causes subsequent loss of other species.
3 D

## p89

B

# INDEX